Space Technologies and Climate Change

IMPLICATIONS FOR WATER MANAGEMENT, MARINE RESOURCES AND MARITIME TRANSPORT

OECD

ORGANISATION FOR ECONOMIC CO-OPERATION AND DEVELOPMENT

The OECD is a unique forum where the governments of 30 democracies work together to address the economic, social and environmental challenges of globalisation. The OECD is also at the forefront of efforts to understand and to help governments respond to new developments and concerns, such as corporate governance, the information economy and the challenges of an ageing population. The Organisation provides a setting where governments can compare policy experiences, seek answers to common problems, identify good practice and work to co-ordinate domestic and international policies.

The OECD member countries are: Australia, Austria, Belgium, Canada, the Czech Republic, Denmark, Finland, France, Germany, Greece, Hungary, Iceland, Ireland, Italy, Japan, Korea, Luxembourg, Mexico, the Netherlands, New Zealand, Norway, Poland, Portugal, the Slovak Republic, Spain, Sweden, Switzerland, Turkey, the United Kingdom and the United States. The Commission of the European Communities takes part in the work of the OECD.

OECD Publishing disseminates widely the results of the Organisation's statistics gathering and research on economic, social and environmental issues, as well as the conventions, guidelines and standards agreed by its members.

This work is published on the responsibility of the Secretary-General of the OECD. The opinions expressed and arguments employed herein do not necessarily reflect the official views of the Organisation or of the governments of its member countries.

Foreword

This book examines the contributions that space technologies can make in tackling some of the serious problems posed by climate change. Focusing on examples of water management, marine resources and maritime transport, it sets out the rationale for further developing satellite systems to measure and monitor climate change and help mitigate its consequences. The report underlines the need to consider satellites not just as research and development systems, but as an important component of a critical communication- and information-based infrastructure for modern societies. The tool box for decision makers that concludes the book reviews different methodological options for deciding on investments in space-based earth observation.

This is the fourth publication related to space issues produced by the International Futures Programme (IFP). The role of the IFP, the OECD's strategic foresight group, is to alert the Secretary-General and the Organisation to emerging issues by pinpointing major developments and analysing long-term concerns, so as to help governments map their strategy.

In 2002, in collaboration with the space community, the OECD IFP launched a project to explore how space technologies could contribute to finding solutions to some of the major challenges facing society. Two publications resulted from that in-depth project. Space 2030: The Future of Space Applications explored promising space applications for the 21st century. Space 2030: Tackling Society's Challenges assessed the strengths and weaknesses of the regulatory structures governing space, and formulated a framework that OECD member countries might use in drafting policies to ensure that the potential offered by space is fully realised.

Upon completion of the two-year project, the IFP was strongly encouraged by a number of institutions – especially space-related agencies – to continue exploring the economic dimensions of space infrastructure. February 2006 thus saw the launch of the OECD Global Forum on Space Economics. This is an innovative platform for international dialogue on the social and economic aspects of space activities (see Annex A for a description). One of the Forum's most recent outputs is The Space Economy at a Glance (2007), the first-ever OECD statistical overview of the emerging space economy.

The Forum is supported by contributions from a number of governments and space agencies: ASI (Agenzia Spaziale Italiana, the Italian Space Agency), BNSC (British National Space Centre), CNES (Centre National d'Études Spatiales, the French Space Agency), CSA (Canadian Space Agency), ESA (European Space Agency), NASA

(US National Aeronautics and Space Administration), NOAA (US National Oceanic and Atmospheric Administration), Norwegian Space Centre (Norsk Romsenter), and USGS (United States Geological Survey).

The book makes particular reference to extensive relevant work conducted in different OECD Directorates. Moreover, several Working Group meetings of the OECD Space Forum were organised over the past two years to discuss and exchange information on the case studies. For example, a special day-long session was organised in October 2007; 25 participants from the sectors of marine resources and maritime transport, as well as representatives from the space sector, were invited to feed the Secretariat's reflections. This report also makes reference to a large number of socio-economic studies that have been carried out over the years, but which, though publicly available, are not well known.

In addition to literature surveys, group and bilateral meetings, the IFP Secretariat conducted interviews with space and non-space experts, and attended external conferences. A first output of the case study work was an OECD Space Forum's internal working paper dedicated to fresh water management and the role of space applications (June 2007). The book builds on the report, and extends the case studies' scope to include more general climate change issues, as well as dedicated work on marine resources and maritime transport.

This publication was prepared by Claire Jolly, Policy Analyst in the Secretary-General's Advisory Unit, under the direction and guidance of Barrie Stevens, Unit Deputy Director, and Pierre-Alain Schieb, Head of Futures Projects, all working for the OECD Global Forum on Space Economics. Shabnam Mirsaeedi provided research assistance, Randall Holden edited the report and Anita Gibson provided administrative and editorial assistance. The report also draws on a number of ground-breaking studies relating to climate change carried out by OECD colleagues, notably in the Environment Directorate (ENV), the Fisheries Policy Division, and the Science, Technology and Industry (STI) Directorate. Special thanks go to Shardul Agrawala (ENV), Jan Corfee Morlot (ENV) and Carl-Christian Schmidt (Fisheries Policy Division) for their contributions.

Table of Contents

Tables

Figures

List of Acronyms

AATSR	Advanced Along Track Scanning Radiometer (instrument on board ENVISAT)
AIS	Automatic Identification System
ASAR	Advanced Synthetic Aperture Radar
ATSR-1 and 2	Along Track Scanning Radiometer (instruments respectively on board ERS-1 and ERS-2)
BRIC	Brazil, the Russian Federation, India and China
BRIICS	Brazil, the Russian Federation, India, Indonesia, China and South Africa
CNES	Centre National d'Etudes Spatiales
CZCS	Coastal Zone Colour Scanner (instrument on Nimbus-7)
DMSP	US Defense Meteorological Satellites Programme
DORIS	Doppler Orbitography by Radiopositioning Integrated on Satellite (instrument on board TOPEX/Poseidon, Jason-1, ENVISAT and the Spot satellites)
DSC	Digital Selective Calling
EEZ	Exclusive Economic Zone
ENVISAT	ENVIronment SATellite
EPIRB	Emergency Position Indicating Radio Beacon
ERS-1 and 2	European Remote Sensing Satellites
ESA	European Space Agency
EUR	Euro (currency of European Union)
FAO	Food and Agriculture Organization
GDP	Gross domestic product
GEOSS	Global Earth Observation System of Systems
GHG	Greenhouse gases
GMDSS	Global Maritime Distress and Safety System
GMES	Global Monitoring for Environment and Security
GOES	Geostationary operational environmental satellites
GOME	Global Ozone Monitoring Experiment (instrument on board ERS-2)
GOMOS	Global ozone measurement by the occultation of stars (instrument on board ESA's ENVISAT satellite)
GOOS	Global ocean observing system
GSE	GMES Services Element

IFREMER	Institut français de recherche pour l'exploitation de la mer
IMAGE	Integrated Model to Assess the Global Environment
IMO	International Maritime Organization
IMSO	International Mobile Satellite Organization
IOC	Intergovernmental Oceanographic Commission of UNESCO
IOOS	Integrated Ocean Observing System
IPCC	Intergovernmental Panel on Climate Change
ISPS	International Ship and Port Facility Security Code
ISRO	Indian Space Research Organisation
ITU	International Telecommunication Union
LANDSAT	LAND observation SATellite
MARS	Monitoring agriculture by remote sensing
MERIS	Medium resolution imaging spectrometer [per MODIS]
MIPAS	Michelson Interferometer for Passive Atmospheric Sounding
MODIS	Moderate resolution imaging spectrometer (instrument on board NASA's Terra and Aqua satellites)
MSR	Maritime search and rescue
MWR	Microwave radiometer
NEXRAD	Next generation radar meteorological stations
NOAA	National Oceanic Atmospheric Administration
NOPP	National Oceanographic Partnership Program
NRT	Near-real-time
OECD	Organisation for Economic Co-operation and Development
POES	Polar operational environmental satellite
ROW	Rest of the world
SAR	Search and rescue
SAR	Synthetic aperture radar satellite
SART	Search and rescue radar transponder
SCIAMACHY	Scanning imaging absorption spectrometer for atmospheric cartography
SOLAS	International Convention on Safety of Life at Sea and its amendments
SSAS	Ship security alert system
SSH	Sea surface height
SST	Sea surface temperature
UNCTAD	United Nations Conference on Trade and Development
UNEP	United Nations Environment Programme
UNESCO	United National Educational, Scientific and Cultural Organization
UNFCCC	United Nations Framework Convention on Climate Change
USCG	US Coast Guard
USD	United States dollar

VMS	Vessel monitoring system
WHO	World Health Organization
WMO	World Meteorological Organization
WSIS	World Summit on the Information Society
WTO	World Trade Organization

Executive Summary

Climate change is emerging as one of the greatest long-term challenges facing society. In fact, it is a set of challenges: how to broaden our understanding of the planet's climate and the ways it is changing; how to better predict and measure its impacts; how to mitigate some of those impacts as effectively and equitably as possible; and, where mitigation is not viable, how best to adapt to it. What role does space have to play in responding to this complex set of challenges?

The report begins by examining briefly the drivers behind climate change; expected future changes to the atmosphere and to sea levels; the likely increase in extreme weather events linked to climate change; and some of the more extreme phenomena (wild cards) that appear less probable but could have very serious impacts indeed (Chapter 1). Climate change is a modification in long-term weather patterns mainly caused by greenhouse gases, which make the earth warmer by trapping energy in the atmosphere. A warmer earth leads to modifications in rainfall patterns and fresh water availability, rises in sea level, and many different effects on plants, wildlife and human activities. It should be borne in mind that a high degree of uncertainty attaches to the various predictions and the science underlying them – as demonstrated by the long-standing worldwide scientific and political debate on these matters. This underscores the importance of better data, better analysis and better science – both to further our knowledge of climate change and its effects on the natural environment and human activity, and to comprehend better the effects human activity is in turn having on natural resources and climate change itself. It is with regard to this knowledge-enhancing function that space has a vital and often unique role to play.

To better illustrate the interplay among climate change, natural resources and human activity – and the important role (actual and potential) of space tools – the report focuses on three "case studies": fresh water management, marine resources, and maritime transport. It shows how the human population and economic pressures already bearing on fresh water resources, fish stocks, etc. are likely to be further exacerbated by changes in precipitation patterns, rising ocean temperatures and shifts in ocean currents. It also shows how melting ice sheets and likely increases in extreme weather events such as hurricanes will affect shipping routes and maritime traffic more generally, and how growing maritime activity itself will impact increasingly on the health of the oceans (Chapters 2 and 3).

All these trends suggest that a concerted effort will be required to gain a full awareness of the dynamics of this interplay, monitor and measure its effects, and improve the mitigation and management of its consequences. The report details the current state of the art in space applications with respect to their capabilities in providing essential data on climate and ocean variables (cloud cover, water vapour, rainfall, soil moisture, salinity, snow cover, water temperature, etc.); in helping to monitor and control marine areas and maritime activity; and in improving safety at sea (Chapter 4).

A variety of satellites and ground systems are already in place, although many are still being developed as short-term R&D programmes for scientific research. These range from meteorological satellites to telecommunication, navigation and earth observation satellites. Space has become an increasingly important source of information and an essential data-relaying infrastructure, a recourse in places in the world where ground-based monitoring systems are not deployable. A number of scientific discoveries about climate change have also been made thanks to space-based data. For example, the Franco-American mission Topex-Poseidon has shown through space altimetry that oceans have been rising over the past decade; it has also provided unexpected information for monitoring oceanic phenomena, such as variations in ocean circulation on the level of the El Nino 1997-98 event. In addition, more than thirty "Essential Climate Variables" identified by the United Nations Framework Convention on Climate Change already depend on satellite data. These observations, combined with others from ground-based installations and models, come at a cost that seems to be increasing as the number and length of missions grow: the scientific data provided by those missions, existing and planned, are not satisfying the demand of a growing number of scientists and operational users.

What about the future? The demands on resource sustainability, on safety, on transport efficiency, etc. are growing quickly. Can the space economy deliver the research, technology and management to keep pace with so much rapid change? New space systems and more applications are in the pipeline or already coming on stream (Chapter 5). Moreover, substantial investments in earth observation and meteorological satellite systems will be needed over the next ten years or so, and considerable replacement and network expansion investments will be required thereafter (to 2025-30 and beyond). Rough estimates in the report put total investment needs for earth observation and meteorological space-based networks at around USD 40 billion in the next 15-20 years.

And it cannot be taken for granted that such levels of investment in space systems will automatically be forthcoming. They are subject to difficult, often contentious political, technical and economic decision-making processes. What will be required to ensure that adequate levels of investment in space systems are ultimately secured?

The answer, first and foremost, is a sound tool box to help policy makers arrive at investment decisions based on good data and analysis (Chapter 6). Valid criteria can come from various sources. There are to begin with the traditional methods of identifying the contributions satellites can make, and comparing these to the costs that their construction and operation incur. A set of examples is provided in the report, which serves to illustrate those benefits and costs. Although figures in the many studies conducted so far on space applications may be subject to caveats, there are strong indications that satellite monitoring and communications contribute clear qualitative and quantitative advantages, as long they are integrated in larger information systems. In terms of disaster prevention and management for example, especially in the case of floods, cost avoidance remains a significant positive return on investments, and may well yet be underestimated (*e.g.* in terms of lives saved, reduced damages to property). With regard to cost-efficiency, the possibility of accessing information and communicating anywhere in the world is another positive aspect. An example, and sign of the growing reliance on space observations, is the International Charter for Space and Major Disasters. Maintained by nine space agencies that agreed to supply their respective optical and radar imagery in times of natural disasters, the Charter was activated 175 times in the last eight years. The latest activation was made in May 2008, following the earthquake in Sichuan province: China, a Charter member, used data as inputs to elaborate maps for rescuers. In terms of commercial fishing operations and maritime transport, several studies point to improved productivity, particularly due to GPS plotters. The transit time savings for commercial ship routing is also an essential economic benefit, based on improved weather-related information and real-time communications at sea.

The lack of quantifiable aggregated benefits from the deployment of space-based systems, coupled with the sheer unpredictability of many future events and their outcomes, clearly complicates major investment decisions based on cost-benefit analysis. In light of this, it can be argued that policy makers need to explore new pathways to reaching decisions.

Indeed there are perhaps other, more novel approaches to be explored that could prove a useful support in the decision-making process. One is to consider space tools as infrastructures and compare the relative levels of investment with those required for terrestrial infrastructures (roads, power, rail, etc.). By way of illustration, this report focuses on the earth observation and meteorological satellite infrastructure, which plays a key role in climate monitoring. The study suggests that the overall cost of setting up such a system – including both R&D and operational satellites – is not unduly high, nor are the rates of annual investment to maintain and expand the space infrastructure and its related networks. Another novel angle could be to assess the operational usefulness of data by taking a risk management approach to space-based infrastructure. Interesting parallels can be drawn with the significant role of economic information or weather risk insurance packages. Weather is a major determinant of earnings performance for entire economic sectors (*e.g.*, utilities). The US Department of Transportation estimates that weather-related delays in air transport cost passengers USD 10 billion in lost time and productivity each year in the United States alone. On an even larger scale, systematic climate monitoring may become an essential tool for governments to hedge the risks associated with climate change and unsustainable resources management (in fisheries for example, where many species are facing extinction).

In conclusion, policy makers can create for themselves and the populations they serve the opportunities to learn more about the complexities of climate change, and so benefit from early warning in order to mitigate and better manage potential impacts. As a possible way ahead for earth observation infrastructure in particular, more attention should be paid to building on major decades-long national and international efforts to develop and sustain operational satellite meteorology. Many satellites from different countries already form the space-based Global Observing System (GOS) that the World Meteorological Organisation co-ordinates; GOS makes key weather information available on a daily basis. For climate monitoring to develop fully as a routine activity, with long-term continuity of measurements and the attendant socio-economic benefits, institutions will increasingly have to share in the effort and provide adequate support to agencies responsible for satellite R&D activities and the operational weather agencies, as they monitor the state of the planet and inherit new climate-related tasks.

ISBN 978-92-64-05413-4
Space Technologies and Climate Change
Implications for Water Management, Marine Resources
and Maritime Transport
© OECD 2008

Chapter 1

Climate Change: Trends and Outlook

Climate change is emerging as one of the greatest long-term challenges facing society. Its impacts have already been felt worldwide, and much worse is looming. Compounding the problem is the uncertainty that surrounds the measurement of climate change and even the phenomenon itself, notably with respect to variations observed in the oceans, seasonal precipitations and extreme weather patterns. This chapter presents some of the main drivers of climate change, reviews trends in extreme weather events, describes possible wild cards for the future, and identifies key challenges in understanding, measuring and forecasting climate change.

Introduction to climate change

By definition, climate change is a modification in long-term weather patterns, likely caused by greenhouse gases that make the earth warmer by trapping energy in the atmosphere. A warmer earth leads to variations in rainfall patterns and fresh water resources, rises in sea level, and a wide range of impacts on plants, wildlife and human activities. It should be borne in mind, however, that a high degree of uncertainty is linked to predictions of this kind and the scientific basis underlying them, as demonstrated by the decades-old worldwide scientific and political debate on these matters. Better data, better analysis and better science are needed – both to improve our understanding of climate change and its impacts on the natural environment and human activity, and to comprehend better how in turn human activity is impacting on natural resources and climate change itself.

Main drivers of climate change

Although much debate continues over the mechanisms, according to the Fourth Assessment of the Intergovernmental Panel on Climate Change (IPCC, 2007), the warming of the climate system is today "unequivocal". This has been established by the increases in global air and ocean temperatures observed, widespread melting of snow and ice, and the rise in global mean sea level.

Eleven of the last twelve years (1995 to 2006) are ranked among the warmest since 1850 (IPCC, 2007). The temperature increase of 0.73 degrees Celsius in the last 100 years has been one signal, but so has the accelerating rate of this increase in temperatures: over the last 50 years, it is almost twice as high as that of the previous 50 (OECD, 2008). This change has been particularly apparent over the northern hemisphere's large landmasses (IPCC, 2007), and has also been noted in sea temperatures. "[O]bservations since 1961 show that the average temperature of the global ocean has increased to depths of at least 3000 m and that the ocean has been taking up over 80% of the heat being added to the climate system" (IPCC, 2007).

A number of factors driving this warming have been identified; one is unsustainable land use policies, such as large-scale deforestation (OECD, 2008). However, greenhouse gases are thought to be the main human-induced driver of climate change. These gases include carbon dioxide, methane and nitrous oxide; they come mainly from energy production facilities and transport activities. Among the gases, carbon dioxide is the dominant contributor to climate change (75% of global emissions), increasing from 278 parts per million (ppm) in pre-industrial times to 379 ppm in 2005 (UNEP/GRID Arendal, 2007). Carbon dioxide concentration is rising mostly as a result of fossil fuel burning (American Meteorological Society, 2008). The environment naturally takes up some carbon dioxide and balances it with the atmosphere and the oceans. In the process, the uptake of CO_2 by seawater leads to the increase of the hydrogen ion concentration in the ocean, decreasing ocean acidity levels and thus raising its pH levels. This natural carbon cycle has been overburdened in the post-industrial age, leading to an excess of carbon dioxide in both air and seas. The result is ocean acidification – a decrease in the pH levels in the earth's oceans caused by their intake of anthropogenic carbon dioxide from the atmosphere. Projections to 2100 estimate a decrease of global ocean pH levels of 0.14% to 0.35%, compared with a reduction of 0.1% since the industrial revolution. Some of the many expected impacts will be described in the next two chapters.

Climate change and extreme weather events

Statistics from a number of insurance and reinsurance companies confirm an increase in the number and cost of natural and man-made catastrophes over the past decades. Economic losses from these catastrophes worldwide exceeded USD 70 billion in 2007 (Swiss RE, 2008); more than 20 000 people lost their lives. In the aftermath, property insurers were hit by claims totalling USD 28 billion. A decade comparison of disasters' costs since 1960 is shown in Figure 1.1 (UNEP/GRID Arendal, 2007). During the period 1980-2006, the United States alone experienced 70 weather-related disasters that each incurred overall damages exceeding USD 1 billion at the time of the event (US Climate Change Science Program, 2008).

Figure 1.1. **Great natural catastrophes: overall economic losses and insured losses, 1950-2007**

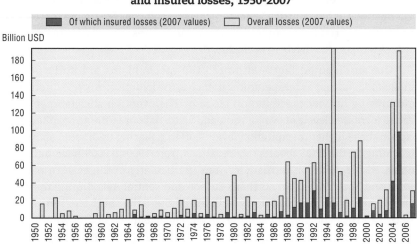

Source: Munich RE, 2008.

Around 90% of all natural disasters are hydrometeorological, and are caused by hurricanes and floods (United Nations, 2003). Hurricane Katrina cost insurers around USD 50 billion, but the total economic damage it inflicted was estimated to be as high as USD 200 billion. The reasons for rising cost trends are manifold and encompass population increase, higher concentrations of people and economic values in vast urbanised areas, and the development of highly exposed regions, especially in coastal areas (i.e. increased risks of flooding and hurricanes).

The link between the rise in storms' frequency and severity and climate change is still controversial. While some scientists and policy makers do not accept a direct link between the two, the most recent IPCC report has rated the correlation as "more likely than not" (IPCC, 2007), since many global and local environmental cycles seem to be undergoing dramatic changes. As an example, there has been an increase recently in the number and intensity of tropical storms and hurricanes in the North Atlantic. Tropical storms are formed when local sea surface temperatures are above 26.5 C° and the resulting ocean evaporation creates humidity that generates thunderstorms. The higher the temperature of the local sea surface, the more likely it is to generate heavy rains and winds. Recent studies indicate that from 1850 to 1990, the overall average number of tropical storms was about ten in the region, including five hurricanes. Since 1995 the ten-year average has risen dramatically; for 1997-2006 the average was about 14 tropical storms, including about eight hurricanes (Pew Center on Global Climate Change, Hurricanes and Global Warming, 2008). On the other hand, some studies (using proxy records of vertical wind shear and sea surface temperature) point out that the average frequency of major hurricanes decreased gradually from the 1960s until the early 1990s, reaching anomalously low values during the 1970s and 1980s. Given this history, the recent phase of enhanced hurricane activity would not be unusual compared to other periods of high hurricane activity on record. It could therefore represent a recovery to normal activity, rather than a direct response to increasing sea surface temperature (Nyberg et al., 2007).

One undisputed fact – although the mechanisms are still not well understood – is that as temperatures increase, the hydrological cycle intensifies, encouraging heavier precipitation globally. The overall higher temperatures lead to increased levels of evaporation and water vapour in the atmosphere that explain these higher precipitation levels – often replacing traditional snowfalls. According to the IPCC reports, observations from 1900 to 2005 have indicated increased precipitation in many regions, particularly in the eastern parts of North and South America, northern Europe and central Asia. During the same period, however, precipitation declined in the Sahel and Mediterranean, southern Africa and parts of southern Asia. The likelihood of drought seems to have thus increased substantially since the 1970s (IPCC, 2007, Topic 2). As globalisation progresses, large-scale disasters in a given country could have major economic repercussions in its neighbouring countries, but also with its worldwide commercial partners. As the underlying factors for the observed loss trend remain unchanged, a further increase in losses from natural disasters seems inevitable – a situation that will increasingly require better monitoring and improved mitigation strategies.

Outlook: unpleasant surprises for the near to medium-term future?

A number of countries have made significant efforts to take into account climate change in their economic policies. These are not sufficient, however, to ensure the environmental sustainability of economic development (OECD, 2008). Without further policies, for example:

● Global emissions of greenhouse gases would grow by a further 37% by 2030, and by 52% to 2050. This could result in an increase in global temperature over pre-industrial levels in the range of 1.7-2.4° Celsius by 2050, leading to increased heat waves, droughts, storms and floods, and the attendant severe damage to key infrastructure and crops.

● Water scarcity would worsen due to unsustainable use and management of the resource as well as climate change. The number of people living in areas affected by severe water stress is expected to increase by another 1 billion reaching over 3.9 billion (OECD, 2008).

In addition to those trends, there are many risk-laden wild cards linked to climate change that could affect numerous areas and activities, including water, marine resources and maritime transport.

Sea level rise

The IPCC estimates that the global average sea level will rise from 0.18 to 0.59 metres in the next century (IPCC, 2007). The range reflects uncertainty about global temperature projections and how rapidly ice sheets will melt or slide into the ocean in response to the warmer temperatures. But recent studies suggest that the IPCC's forecast may be too conservative. For instance, many of the dynamic processes of ice sheet activity are not included in some projections, which thus ignore the potential impact of their melting (Rohling et al., 2008). According to Overpeck et al. (2006), polar warming by the year 2100 may reach levels similar to those of 130 000 to 127 000 years ago that were associated with sea levels several metres above modern levels. Many unknowns remain and more data and research are needed to refine sea level projections (see Figure 1.2).

Even with conservative estimates, the impacts of sea level rises are numerous; they include land area loss, people displacement, ecosystem loss and changes, economic value loss, important human infrastructure loss,

Figure 1.2. **Observed and projected global average sea level rise, 1800-2100**

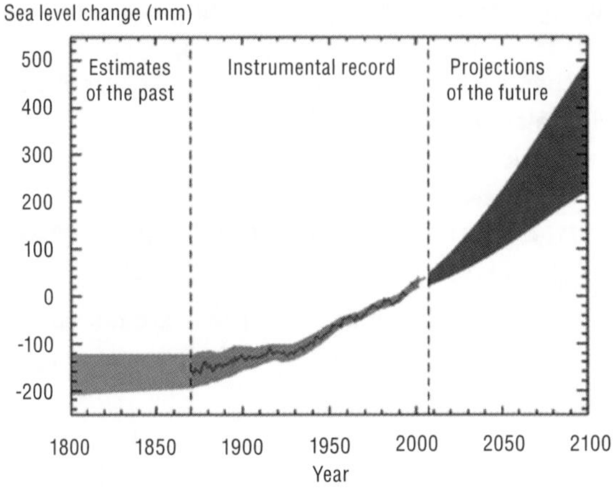

Sea level change (mm)

Source: Solomon *et al.*, 2007.

cultural loss and of course important adaptation costs (OECD, 2006). The world's coastlines stretch to 1.63 million kilometres, and nearly half (46%) of them are located in OECD countries (*e.g.* the long coastlines of Canada, the United States, Mexico and Australia). Most importantly, coastal zones (areas within 100 kilometres of a shoreline and 100 kilometres of sea level) are home to 1.2 billion people, or a fifth of the world's population. Overall, average population density in coastal zones is three times higher than the world average, and in recent decades the overall growth of coastal populations has outstripped that of inland populations (Crawford Heitzmann, 2006). Sea level rise is already having significant effects on coastal communities and industries, islands, river deltas and harbour areas worldwide. It could increase the salinity of bays and estuaries, and coastal erosion. As an example, a 50 centimetre rise in sea level will typically cause a shoreward retreat of coastline of 50 metres if the land is relatively flat (like most coastal plains), causing substantial economic, social, and environmental impacts.

Although the sea level is not rising uniformly around the world, future trends indicate an overall increase in both population and asset exposure to the effects of the rise (see Figures 1.3 and 1.4). Ten large cities are already exposed to present-day extreme sea levels and could see worsening impacts: Tokyo, New York, Shanghai, Kolkata, Dhaka, Osaka, Mumbai, Guangzhou, Shenzhen and Miami. Rapidly growing cities in developing countries in Asia, Africa and to a lesser extent Latin America are also expected to be increasingly exposed in the next decades (Nicholls *et al.*, 2007). In most projections, it is assumed that south and east Asia will be the most vulnerable because of the large coastal populations in low-lying areas, such as Vietnam, Bangladesh and parts of China and India. Millions will also be at risk around the coastline of

Africa, particularly in the Nile Delta and along the west coast. Small island states in the Caribbean, and in the Indian and Pacific Oceans (*e.g.* Micronesia and French Polynesia, the Maldives, Tuvalu) are acutely threatened because of their high concentrations of development along the coast, while in the Caribbean more than half the population lives within 1.5 kilometres of the shoreline. Some estimates suggest that 150–200 million people may become permanently displaced by the middle of the century due to rising sea levels, more frequent floods, and more intense droughts (Stern, 2007).

Figure 1.3. **Assets exposed to sea level rise, storm surge and subsidence by 2070, by country**

Total estimated exposure = USD 35 000 billion

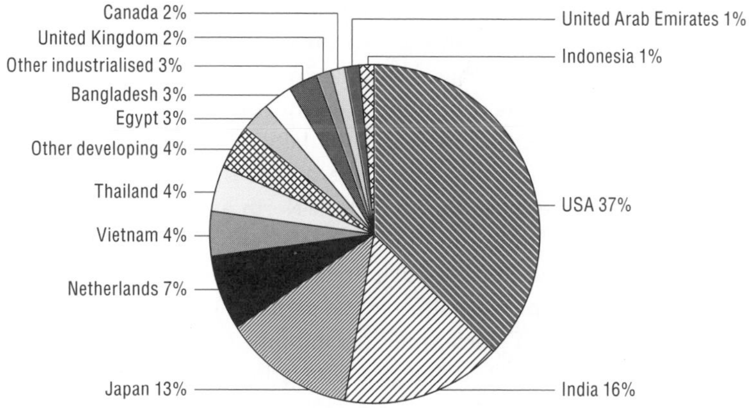

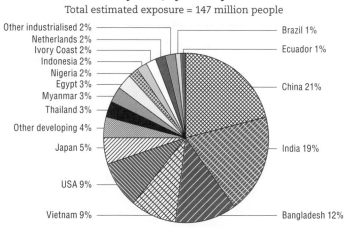

Canada 2%
United Kingdom 2%
Other industrialised 3%
Bangladesh 3%
Egypt 3%
Other developing 4%
Thailand 4%
Vietnam 4%
Netherlands 7%
Japan 13%
United Arab Emirates 1%
Indonesia 1%
USA 37%
India 16%

Source: Nicholls *et al.*, 2007

Figure 1.4. **Population exposed to sea level rise, storm surge and subsidence by 2070, by country**

Total estimated exposure = 147 million people

Other industrialised 2%
Netherlands 2%
Ivory Coast 2%
Indonesia 2%
Nigeria 2%
Egypt 3%
Myanmar 3%
Thailand 3%
Other developing 4%
Japan 5%
USA 9%
Vietnam 9%
Brazil 1%
Ecuador 1%
China 21%
India 19%
Bangladesh 12%

Source: Nicholls *et al.*, 2007.

Weakening of oceanic currents

Several oceanic circulation systems are naturally variable, such as the North Atlantic Oscillation (NAO), the Atlantic Meridional Overturning Circulation (AMOC) and El Niño Southern Oscillation (ENSO); these affect the world climate. They are not yet perfectly understood, but all of them seem to respond to the increasing climate change variations. The following paragraphs will highlight trends observed in two of the systems.

a) The Atlantic Meridional Overturning Circulation (AMOC)

The possible consequences of a weakening Atlantic Meridional Overturning Circulation (often referred to as the Gulf Stream) are far-reaching. The Gulf Stream consists of a large-scale current that flows in the Atlantic, transporting warm waters northward. As the current pushes into the North Atlantic the surface water is cooled by winds from the Arctic; it becomes saltier and denser, and sinks in the cold high northern latitudes. The cold water then returns to the Southern Ocean at a depth of between 1 500 and 3 500 metres. Although there is still limited data coverage, and so the evidence is not totally conclusive, recent studies have shown that the North Atlantic Ocean circulation system seems to have weakened in the late 1990s, compared to the 1970s and 1980s (Häkkinen and Rhines, 2004; Bryden, Longworth and Cunningham, 2005). The oceanic current is sensitive to the paths of winter storms and to the buoyant fresh water from glacial melting and precipitation, all of which are experiencing changes. In that context, the circulation system gives off heat as it moves north, creating warmer, milder European winters.

Many climate models suggest that a total shutdown of the AMOC within the 21st century is very unlikely. Research is ongoing, however, to study the effects of even partial shutdowns: a decline in the strength of this large-scale current, associated with increasing greenhouse gases, could exacerbate warming in and around the North Atlantic (IPCC, 2007, Chapter 10). Future projections may well show that once the radiative forcing is held fixed, AMOC is re-established to a degree similar to that of the present day. But this too has consequences, as the partial or complete re-establishment of the AMOC is slow and causes additional warming in and around the North Atlantic. The synthesis of different climatologists' views and models in Zickfeld *et al.* (2007) suggests for example that a regional shutdown of deep-water formation in the Labrador or in the Greenland–Iceland–Norwegian Seas, or even a total cessation of the circulation, could lead to a marked cooling of several degrees in the North Atlantic region and an increase in sea level of up to 1 metre in magnitude. Indirect global effects could include a shift of the Inter-tropical Convergence Zone, a belt of low pressure girdling earth at the equator, and a warming of the Southern Ocean.

b) El Niño Southern Oscillation (ENSO)

The El Niño Southern Oscillation (ENSO) occurrence is a year-to-year fluctuation of the global climate system arising from large-scale interactions between the ocean and the atmosphere, with widespread environmental and meteorological impacts. As with the AMOC, there is some uncertainty associated with ENSO's links to climate change, and many intertwined issues regarding ENSO dynamics, impacts, forecasting and applications remain unresolved (McPhaden, Zebiak and Glantz, 2006; Meehl et al., 2006).

El Niño and La Niña phenomena occur regularly during the April-June period and reach their maximum strength during December-February.* These events typically last 9-12 months, although they can sometimes last up to two years (International Research Institute for Climate and Society, 2008). El Niño and La Niña shift the probability of droughts, floods, heat waves and extreme weather events in many parts of the globe, including the Indian subcontinent, eastern and southern Africa, South America, and southern North America. During El Niño, warm water currents flow towards the coasts of Chile and Peru from the western Pacific, causing warm and wet weather in South America. In contrast, the western Pacific regions can suffer extensive droughts. The reverse happens during La Niña.

ENSO conditions have occurred every two to seven years, but the rapid succession of El Niño events in the 1990s and its strength in 1997-98 stimulated debate as to whether climate change increases the intensity or frequency of El Niño episodes. The first successful El Niño prediction occurred in 1986; forecast systems development was motivated by ENSO's socio-economic impacts. The predictability probabilities and the models are still under development as much remains unknown about ENSO phenomena, but current systems provide forecasts of six- to nine-month lead times (McPhaden, 2007).

Instability of the continental ice sheets

The melting of polar ice sheets – along with the ensuing sea level rise – is one of the largest potential threats of future climate change. As noted above, sea level rise is already agreed by the international scientific community. One wild card concerns the speed at which water will rise over the next decades, producing major socio-economic consequences.

The Arctic may be the location of the most rapid and dramatic climate changes during the 21st century, with major ramifications for mid-latitude climate (see Box 1.1). The latest Arctic Climate Impact Assessment studies (ACIA, 2005) have identified a number of severe potential impacts of Arctic warming on society. Three hundred scientists participated in the study over a

---

* El Niño is the warm phase of ENSO, whereas La Niña is the cold phase.

Box 1.1. **Changes in the Arctic region**

The reduction in sea ice observed in the Arctic Ocean in the 20th century occurred against a backdrop of significant annual and multi-annual atmospheric variability. In 2005, when temperatures in the Arctic region north of the 65th parallel were the highest since recordkeeping began in the mid-19th century, the area of sea ice in the northern hemisphere in September decreased to its lowest level since satellite (and hence more accurate) observation began. The layer of Atlantic water in the Arctic Ocean has increased and become warmer. Over the past decade, land snow cover in the northern hemisphere has been reduced. There are some indications that precipitation in the Arctic region increased in the 20th century. Also, the mass balance of ice in the northern hemisphere was negative, and warming was observed in numerous permafrost areas.

There has been an increase in the annual flow of rivers entering the Arctic Ocean, and its seasonal redistribution. In particular, over the past 20 to 25 years, the main change in the seasonal flow of rivers in the Arctic Ocean basin has been the marked increase in their water levels in winter. The total annual flow of the six largest rivers in Russia (Pechera, Ob, Yenisei, Lena, Yana and Indigirka) from 1936, when records began, until 1999 has increased by 7%, and is still rising. Whereas over the entire 20th century the temperature worldwide rose by 0.6 degrees Celsius, in the Arctic region it rose by 5 degrees. According to scientists, this trend is set to continue, bringing with it the threat of more natural disasters.

Source: Adapted from Grachev, 2008.

span of three years and agreed that changes in air temperature, precipitation, river discharge, sea ice, permafrost, glaciers and sea level are occurring, and further changes are expected over the next decades. A recent report by the US National Oceanic and Atmospheric Administration provided an update to some of the records of physical processes discussed in the Arctic Climate Impact Assessment, showing that sea ice extent continued to decrease. A minimum in sea ice extent was again observed in September 2007, marking an unprecedented series since the beginning of direct satellite observations in 1979 (National Snow and Ice Data Center, 2007).

As mentioned by the National Snow and Ice Data Center (2007), despite considerable year-to-year variability, significant negative trends are apparent in both maximum and minimum sea ice extents, with a rate of decrease of –2.9% per decade for March and –10.5% per decade for September. The differences in extent are calculated as anomalies compared to the 1979-2000 average, which was 7 million square kilometres in September, and 15 million square kilometres in March. September 2007 presented a record low, with only 4.3 million square kilometres. Other changes observed using data from satellites and *in situ* missions include a warming of the Atlantic water in the Arctic Ocean, increased precipitation in the Arctic regions, and higher river discharge into the Arctic Ocean (Box 1.1).

Annual temperatures in the Antarctic region have also been increasing, and ice shelves on the fringes of the peninsula seem to be disintegrating. Recent findings have shown that snow accumulation may have doubled in the western Antarctic Peninsula since 1850 and especially over the past 50 years, the period of observational record (Thomas *et al.*, 2008). This could be caused by shifts in the circulation patterns of the Southern Annular Mode (SAM), a dominant cycle of atmospheric variability in the southern hemisphere. More warm moist air seems to reach the region, provoking an increase in snow accumulation as this air cools. Much uncertainty remains however; more data and modelling are required to understand those complex processes.

Uncertainties and key assumptions

As this chapter has shown, addressing climate change entails meeting a set of important challenges: how to better understand the planet's climate and the ways it is changing; how to better predict and measure its impacts; how to mitigate some of those impacts as effectively as possible; and, where mitigation is not viable, how best to adapt to them.

Responding to those challenges calls for a major effort on two fronts:

- First, closing the gaps in our knowledge. A number of climate-related sciences are progressing rapidly and more information is becoming available, but data to help better understand atmosphere-, land- and ocean-related processes are still lacking in many instances.

- Second, reducing uncertainty surrounding future projections (Box 1.2). Advances in climatology and modelling techniques are key in this respect and will require continued improvement in the collection, range and quality of climate-related data.

Box 1.2. **Selected uncertainties and assumptions concerning climate change projections**

Projections of climate change depend on a number of parameters, all of which reflect future uncertainty:

- *Climate sensitivity* – This parameter characterises how global temperatures respond to a doubling of CO_2 concentrations. In its 2007 report the IPCC noted that climate sensitivity is likely to be in the range of 2-4.5°C with a "best estimate" of 3.0°C. It is very unlikely to be below 1.5°C, and values substantially higher than 4.5°C cannot be excluded.

- *Abrupt changes and surprises* – The OECD Outlook Baseline (OECD, 2008) assumes a linear response to increasing concentrations of greenhouse gases. There is, however, evidence from the palaeoclimatic record that the earth's systems have undergone rapid changes in the past, and that these could occur in the future.

- *Probability of outcomes, risk assessment* – Given these and other uncertainties, probabilistic assessment is increasingly used to give policy makers an idea of the likelihood of achieving identified targets. For example in the case of a 2°C target, it was estimated that a 650 ppm CO_2 eq concentration level would offer only a 0-18% probability of success. This places climate change in a risk assessment and management framework.

- *Adaptation* – Human systems are likely to respond to climate change through adaptation, while ecological systems are likely to find it more difficult to adapt. The faster global warming occurs, the more difficult and limited adaptation will be. Most current studies of climate change impacts recognise the need to consider adaptation, but few modelling studies integrate adaptation comprehensively into quantitative analyses.

Source: Based on OECD, 2008.

Bibliography

ACIA (Arctic Council and the International Arctic Science Committee) (2005), *Arctic Climate Impact Assessment Studies*, Cambridge University Press.

American Meteorological Society (2008), "Climate Change: An Information Statement of the American Meteorological Society (Adopted by AMS Council on 1 February 2007)", *Bulletin of the American Meteorological Society*, 88.

Bryden, Harry L., Hannah R. Longworth and Stuart A. Cunningham (2005), "Slowing of the Atlantic Meridional Overturning Circulation at 25°-N", *Nature*, 438, pp. 655-657.

CCSP (Climate Change Science Program) (2008), *Weather and Climate Extremes in a Changing Climate, Regions of Focus: North America, Hawaii, Caribbean, and US Pacific Islands*, US Climate Change Science Program, Synthesis and Assessment Product 3.3, June.

Crawford Heitzmann, Martha (2006), "Don't Forget the Coastal Waters!", *OECD Observer*, March.

Grachev, Vladimir (2008), *Protection of the environment in the Arctic Region*, Report to the Committee on the Environment, Agriculture and Local and Regional Affairs, European Parliament, Doc. 11477, January.

Häkkinen, Sirpa and Peter B. Rhines (2004), "Decline of Subpolar North Atlantic Circulation During the 1990s", *Science*, 304, 23 April, pp. 555-559.

International Research Institute for Climate and Society (2008), ENSO Information Website, <*http://iri.columbia.edu/climate/ENSO/*>, accessed 20 June 2008.

IPCC (Intergovernmental Panel on Climate Change) (2007), *Climate Change 2007 Synthesis Report: An Assessment of the Intergovernmental Panel on Climate Change*, Valencia, Spain, 17 November.

McPhaden, Michael J., Stephen E. Zebiak and Michael H. Glantz (2006), "ENSO as an Integrating Concept in Earth Science", *Science*, Vol. 314, No. 5806, 15 December.

McPhaden, Michael J. (2007), "Lessons Learned from El Niño and La Niña Monitoring", Presentation during OECD Space Forum Working Group Meeting, 19 October.

Meehl, Gerald A. *et al.* (2006), "Climate Change Projections for the Twenty-First Century and Climate Change Commitment in the CCSM3", *Journal of Climate*, Vol. 19, Issue 11, June, pp. 2597–2616.

Munich RE (2008), *Topics Geo Natural Catastrophes 2007: Analyses, Assessments, Positions*, Knowledge Series, Zurich.

Nicholls, Robert J. *et al.* (2007), "Ranking Port Cities with High Exposure and Vulnerability to Climate Extremes Exposure Estimates", Environment Working Papers No. 1, ENV/WKP(2007)1, OECD, Paris.

NRC (National Research Council) (2006), *Surface Temperature Reconstructions for the Last 2,000 Years*, National Academies Press, Washington, DC.

NSIDC (National Snow and Ice Data Center) (2007), Website, <*http://nsidc.org/*>, accessed 20 December 2007.

Nyberg, Johan *et al.* (2007), "Low Atlantic Hurricane Activity in the 1970s and 1980s Compared to the Past 270 years", *Nature*, Vol. 447, June, pp. 698-701.

OECD (2006), "Metrics for Assessing the Economic Benefits of Climate Change Policies: Sea Level Rise", Working Party on Global and Structural Policies, 30-31 March.

OECD (2008), *Environmental Outlook to 2030*, OECD, Paris.

Overpeck, Jonathan T. *et al.* (2006), "Paleoclimatic Evidence for Future Ice-sheet Instability and Rapid Sea-level Rise", *Science*, Vol. 24, March.

Pew Center on Global Climate Change (2008), Hurricanes and Global Warming Website: <*www.pewclimate.org/*>, accessed 20 June.

Rohling, Eelco *et al.* (2008), "High Rates of Sea-Level Rise During the Last Interglacial Period", *Nature Geoscience*, Vol. 1, pp. 38-42.

Solomon, S. *et al.* (eds.) (2007), *Climate Change 2007: The Physical Science Basis*, Contribution of Working Group I to the Fourth Assessment Report of the Intergovernmental Panel on Climate Change, Cambridge University Press, Cambridge.

Stern, Nicholas (2007), *The Economics of Climate Change: The Stern Review*, Cambridge University Press, Cambridge.

Swiss RE (2008), "Natural Catastrophes and Man-made Disasters in 2007", *Sigma,* No. 1/2008, 22 January.

Thomas, Elizabeth R., Gareth J. Marshall and Joseph R. McConnell (2008), "A Doubling in Snow Accumulation in the Western Antarctic Peninsula since 1850", *Geophysical Research Letters*, 35, L01706.

UNEP/GRID-Arendal (2007), "Trends in Arctic Sea Ice Extent in March (Maximum) and September (Minimum) in the Time Period of 1979–2007", UNEP/GRID-Arendal Maps and Graphics Library, 2007, *<http://maps.grida.no/go/graphic/trends-in-arctic-sea-ice-extent-in-march-maximum-and-september-minimum-in-the-time-period-of-1979-2007>* accessed 23 June 2008.

Zickfeld, Kirsten *et al.* (2007), "Expert Judgements on the Response of the Atlantic Meridional Overturning Circulation to Climate Change", *Journal of Climatic Change*, Vol. 82, Nos. 3-4, June.

ISBN 978-92-64-05413-4
Space Technologies and Climate Change
Implications for Water Management, Marine Resources
and Maritime Transport
© OECD 2008

Chapter 2

Fresh Water Management: Trends and Outlook

Water management has become a key issue for the 21st century, as growing pressures are hindering the delivery of already scarce fresh water to millions of people. Expanding populations, economic growth, pollution and seasonal climatic conditions are all factors behind diminishing water resources. In addition however, a number of effects linked to climate change, such as lengthy droughts and extreme weather events, are worsening the situation. The link between water management and climate change makes monitoring and forecasting the global water cycle increasingly important. This chapter points to some of the key challenges related to the management and delivery of water-related services. The first section focuses on the growing pressures surrounding access to fresh water; the second highlights some of the socio-economic impacts of "water troubles" (such as the lack of water access and costs of water-related disasters); and the last explores some of the interactions between fresh water and climate change.

Introduction to fresh water management

Fresh water is essential for life, but only 2.3% of the earth's water is fresh, and two-thirds of that is permanently frozen. Sustained worldwide access to the scarce resource is further jeopardised by growing natural and man-made pressures. Climate change is one of the possible exacerbating factors: it will increasingly affect where, when, how much and how water falls, and at the same time make water supplies more vulnerable, increase the severity of droughts and flooding events, and further threaten fragile coastal aquifers.

Box 2.1. **The OECD and water management**

The OECD has long had an interest in both the environmental effects of economic and social policies and in the economic and social effects of environmental policies (OECD, 2006a). This interest is reflected in a focus on the cost-effectiveness of water management policies, the efficiency of water resource allocation, and the impact on water resources of sectoral and other economic policies (*e.g.* concerning agriculture or spatial planning). In recent years, the OECD has evaluated the effectiveness of economic instruments for, *inter alia*, water management and water pricing for domestic, industrial and agricultural uses. The Organisation has, together with environment ministers from its member countries, developed an *Environmental Strategy for the First Decade of the 21st Century*, endorsed by ministers of economics and finance in 2001.

This strategy highlights water as a priority for policy action, articulating two key challenges facing the member countries:

● Management of fresh water resources and associated watersheds so as to maintain an adequate supply of fresh water suitable for human use and to support aquatic and other ecosystems.

● The need to protect, restore and prevent deterioration of all bodies of surface water and ground water, to ensure the achievement of water quality objectives in OECD countries.

In adopting the strategy, member countries pledged to undertake national action aimed at meeting these challenges, and agreed on three broad indicators for measuring progress: reduced intensity of water resource use; improved ambient water quality; and a larger share of the population connected to secondary and tertiary wastewater treatment systems. The OECD

> Box 2.1. **The OECD and water management** *(cont.)*
>
> is working closely on water policies with a number of organisations; one is the UN Secretary-General's Advisory Board on Water and Sanitation (UNSGAB, 2006), created in 2004 by Kofi Annan to galvanise global action on these issues as part of international efforts to eradicate poverty and achieve sustainable development. In July 2006, the OECD Secretary-General Angel Gurría and UNSGAB's Acting Chair Uschi Eid signed a Joint Statement that outlines the roles of the two bodies in their effort to encourage increased financing for the water sector and to assist national, regional and local efforts to build the capacity to manage water resources.

The growing pressures on fresh water availability

Although the natural balance between fresh water sources and uses has been largely unaffected for centuries in most countries, the global water resources available are diminishing while populations are increasing, leading to greater scarcity. Sources of fresh water are mainly through direct precipitation, which is usually stored temporarily in natural areas or in man-made reservoirs that hold some 8 000 cubic kilometres. Some 8% of the annual fresh water renewable resource is used, with 26% of evapotranspiration of water and 54% of runoff. In 1999, there were reported to be 261 international trans-boundary basins that together cover 45% of the earth's land surface, encompassing 40% of the world's population and providing 60% of the planet's entire fresh water volume. A total of 145 countries' land areas fall partially or completely within international basins.[1]

In an increasingly tense context, to which must be added the threats to regional and global ecosystems caused by anthropogenic and natural climate change, competition for water is increasing among three major sectors: agriculture, industry and domestic consumption (World Bank, 2003; OECD, 2008). Globally and in most major river basins, the biggest volumes are withdrawn for irrigation purposes; agriculture is responsible for approximately 70% of water use worldwide. Governments must increasingly ensure that farmers use water resources efficiently, and that they are allocated among competing demands in a way that enables farmers to produce food and fibre, minimise pollution and support ecosystems, while meeting social aspirations.

Pursuit of the narrow development goal of increased agricultural productivity has led to the breakdown of many resilient ecosystems, which had also been under the strain of climate change effects. That breakdown, combined with competition from other economic sectors and the need to conserve the integrity of aquatic ecosystems, is now progressively limiting the amount of water available to agriculture. Agriculture, although an essential economic sector, has come under pressure to reduce the level of its negative effects, particularly those associated with the use of fertilisers and pesticides and the wasteful use of water. As Table 2.1 shows, demand for water is set to increase dramatically in the next couple of decades, driven by increased population growth, urbanisation, economic growth and changing climate conditions. By 2025, water withdrawals in developing countries could increase by 27% and by 11% in developed countries.

Table 2.1. **Global water use from 1900 to 2025 (Cubic kilometres)**

Use	1900	1950	1995	2025
Agriculture				
Withdrawal	500	1 100	2 500	3 200
Consumption	300	700	1 750	2 250
Industry				
Withdrawal	40	200	750	1 200
Consumption	5	20	80	170
Domestic				
Withdrawal	20	90	350	600
Consumption	5	15	50	75
Total				
Withdrawal	600	1 400	3 800	5 200
Consumption	300	750	2 100	2 800

Source: Ashley and Cashman, 2006.

It is expected that domestic water use will account for 21% of global demand in 2025, compared with 10% in 1995; industrial use should remain relatively unchanged at around 20%, while agricultural use will decline from 70% in 1995 to 56% in 2025 (OECD, 2006b). Water for industrial use should decline in developed countries, due to the changing nature of industrial activity as well as further moves to reduce unit water usage and increase water productivity. The major drivers will be increasingly stringent environmental regulation and the cost of water. In absolute terms, water consumption in North America and western and central Europe will decline along with that in Japan and Australia; the trend is already well established in Europe. However, given both India's and China's pace of urbanisation, their demand for water in the domestic sector is expected to double by 2030; the increase in industrial demand is likely to be similar. For India to meet the increases, part of the current allocation for agriculture will have to be transferred to the domestic and industrial sectors. However, the problems this will cause will be exacerbated as most domestic growth will be located in water-scarce areas of the country. In China there will be a target of nil growth in water consumption for irrigation – which, coupled with further resource developments, should cater for the increased needs in the irrigated areas.

Figure 2.1. **Reduction in global water availability, 1950-2030**

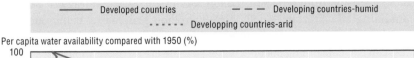

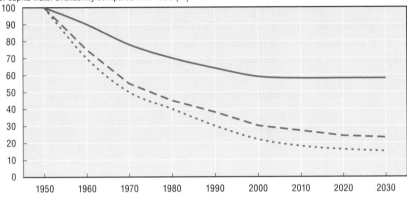

Source: World Bank, 2003.

The increased fragility of ecosystems is demonstrated by the first global study on river flow status, which pointed out the increasing fragmentation of river basins as a result of damming and other flow impediments (United Nations, 2006). This fragmentation is increasingly affecting many countries. Waterfalls, rapids, riparian vegetation and wetlands can all disappear when river flow is altered, by excessive damming in particular. Major local and regional threats to fresh water ecosystems include habitat alteration, land-use change – especially deforestation and agricultural intensification – river fragmentation and flow regulation, water pollution, invasive species and climate change. Failure to address these problems will incur significant social and economic costs and long-term, potentially irreversible effects on biodiversity. Thus there will be a need to monitor – globally, regionally and locally – the changes in water distribution.

With regard to ground water, four countries account (by volume) for more than 60% of its abstraction worldwide: India, China, Pakistan and the United States (see Table 2.2). These countries account for 75% of ground water's agricultural usage. Data are not, however, as reliable as they are for private well abstractions for individual farmers – thus agricultural use is likely to be severely underestimated for many countries like India. Over-abstraction, where withdrawal exceeds recharge, is occurring in a number of Middle Eastern countries and also in Turkey and Mauritania. By 2000, some 25% of Mexico's aquifers were already known to be overexploited in arid parts of the country, and recent droughts have worsened the situation. A similar picture is expected for India and certain regions of China. In India, Pakistan, China and Iran a significant proportion of the population depends on ground water abstractions, through a high number of very small structures. Over-extraction of ground water can lead to saline intrusion from adjacent coastal waters, or other naturally or human-contaminated water. In Bangladesh, 22 million people risk being poisoned by naturally occurring arsenic in ground water. In these areas there may be no alternative to consuming polluted water as there are no other affordable sources, although efforts are under way to find a solution, *e.g.* direct rainwater harvesting. Arsenic is a more widespread problem in ground water and has recently been reported in sources used for supply in Italy, Pakistan, Mexico and China.

Table 2.2. **Ground water abstraction in selected regions**

	Number of structures extracting groundwater (thousands, 2003)	Average extraction per structure (m^3/year)	Population dependent on groundwater (%)
India	19 000-26 000	7 900	55
Pakistan	500	90 000	60
China	3 500	21 500	22
Iran	500	58 000	12
Mexico	70	414 285	5
United States	200	500 000	< 1

Source: Ashley and Cashman, 2006.

The socio-economic risks of inadequate water supplies

As seen earlier, the current trends in fresh water abstraction and the increased fragility of ecosystems, acting in combination with climate change-related effects such as droughts, are producing a rather pessimistic scenario for large segments of the world population. A set of related socio-economic issues is also contributing to the bleak picture. In addition to fresh water's lack of availability, water quality is becoming a daunting challenge. It is estimated that 2 million tonnes of waste is already discharged daily, polluting some 12 000 cubic kilometres of receiving waters (OECD, 2006b).[2]

Although progress has been made in extending safe water and sanitation coverage, further improvement is still needed to meet global targets. This is particularly the case in sub-Saharan Africa. The sanitation target – to halve the proportion of people lacking sanitation – will not be met by 2015 without additional effort. Currently, over 1 billion people in the world are still without access to improved water supplies, and some 2.4 billion people are without access to sanitation. The overall trend in most countries is that improved water supply reduces mortality rates, and the beneficial effects are greater when sanitation is introduced. Worldwide, between 1990 and 2002, about 1.1 billion people had access to improved water sources, while global sanitation coverage rose from 49% (in 1990) to 58% (in 2002).

Table 2.3. **Percentage of population served by water supply and sanitation services, 2002**

Infrastructure		World	Developed countries	Eurasia	Developing regions
Water supply	Urban	95	100	99	92
	Rural	72	94	82	70
	Total	83	98	93	79
Sanitation	Urban	100	100	92	73
	Rural	37	92	65	31
	Total	58	98	83	49

Source: Ashley and Cashman, 2006.

Inadequate drinking water supply and poor water quality and sanitation are therefore still among the main causes of preventable disease and death in the world. While the majority of these deaths occur in developing countries, OECD countries are not immune to outbreaks and fatalities from water-borne disease. Many new scientific and technological developments are helping to make more efficient use of available water resources, to reduce pollutants in water bodies, and to improve treatment of drinking water – but much remains to be done. The value of providing water and sanitation in developing countries is highlighted by the World Health Organization (Hutton *et al.*, 2004; Prüss-Üstün *et al.*, 2008), which sets out *benefit/cost* ratios for that provision (Table 2.4).

Table 2.4. **Benefit/cost ratios for water management interventions in developing regions and Eurasia**

Intervention	Annual benefits in USD millions	Benefit/cost ratio
Halving the proportion of people without access to improved water sources by 2015	18 143	9
Halving the proportion of people without access to improved water sources and sanitation by 2015	84 400	8
Universal access to improved water and sanitation services by 2015	262 879	10
Universal access to improved water and sanitation and water disinfected at the point of use by 2015	344 106	12
Universal access to a regulated piped water supply and sewage connection in house by 2015	555 901	4

Source: Hutton *et al.*, 2004.

Notably, halving the number of people without access to water supplies and sanitation by 2015 would cost an annual USD 11.3 billion in investment, but generate a potential payback in the order of USD 84 billion per year in gains from improved health, lower mortality, higher incomes, etc. That represents a benefit/cost ratio of nearly 8. Conversely, failure to achieve those targets comes at a high price in terms of mortality, sickness, lost income, medical care expenses, etc.

The significant water management efforts of past decades have not been enough to safeguard and restore water quality and aquatic ecosystems, especially with the worsening effects of climate change. The future demand for and consumption of water will be influenced not just by climatic factors but also by policy decisions, the actions of millions of individuals, the type of water infrastructure and services and access to them, and changes in technology. Key decision making is required to protect populations and prevent larger economic losses. That raises the question of whether and to what extent space systems can play a role in helping decision makers handle the issues described in this chapter. The chapters that follow offer an analysis.

Fresh water and climate change

In addition to the pressures identified so far (i.e. demography, economic development), water as a natural resource is influenced by climatic factors (OECD, 2008). As mentioned by Stern (2007), one impact of climate change that will be felt strongly is the change in distribution of water around the world and its seasonal and annual variability.

Figures 2.2 and 2.3, prepared for the OECD *Environmental Outlook* (OECD, 2008), indicate country groupings and geographic areas likely to experience various degrees of water stress by 2030.

Figure 2.2. **People living in areas of water stress, by level of stress**
Millions of people

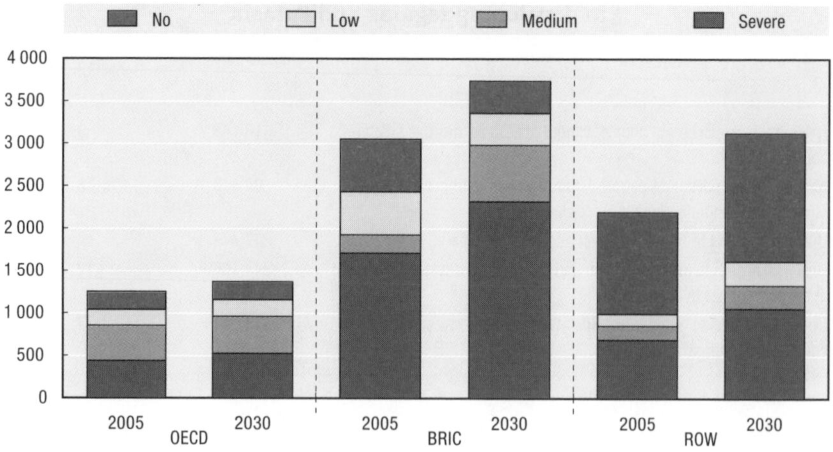

Source: OECD, 2008.

The maps show that medium (light blue) to severe (black) water-stressed areas are set to expand, most notably in Africa and south Asia. More important than the area, however, is the growing number of people depending on water in these areas.

Figure 2.3. **Water stress by major water basin in 2005 and 2030**

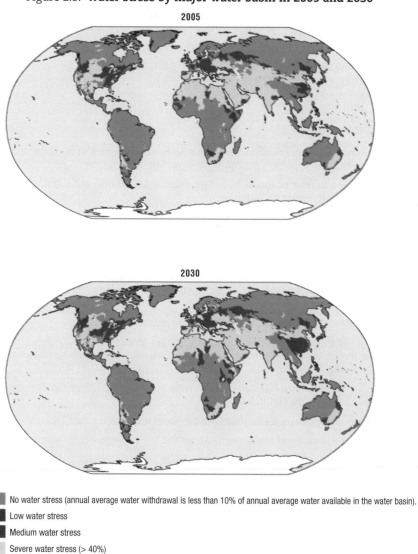

No water stress (annual average water withdrawal is less than 10% of annual average water available in the water basin).

Low water stress

Medium water stress

Severe water stress (> 40%)

Source: OECD, 2008.

The projected change in climate will significantly affect the hydrological cycle, exacerbating in some cases water scarcity and droughts via changes in precipitation patterns and increased rates of evapotranspiration. A number of scenarios estimate that there could be more rain at high latitudes, less rain in the dry subtropics, and uncertain but probably substantial changes in tropical. For example, several climate models predict a decrease of up to 30% in annual runoff in these regions ensuing from a 2°C global temperature rise, and a 40-50% decrease from a rise of 4°C (Warren, 2006). Warmer weather is also likely to translate into an increase in the occurrence and intensity of water quality problems in many regions of the world (*e.g.* harmful algal blooms as surface waters warm, and saltwater intrusion resulting from storm surge and coastal flooding) (IPCC, 2007; OECD, 2006d).

Notes

1. Conflicts over water may increase as diminished water quality or quantity may destabilise regions, especially within trans-boundary basins.

2. As will be seen in later chapters, better monitoring of national and international waters might help lessen pollution.

Bibliography

Ashley, Richard and Adrian Cashman (2006), "The Impacts of Change on the Long-term Future Demand for Water Sector Infrastructure", in *Infrastructure to 2030: Telecom, Land Transport, Water and Electricity*, OECD, Paris.

Hutton, G. and L. Haller (2004), *Evaluation of the costs and benefits of water and sanitation improvements at the global level*, World Health Organization, Geneva.

IPCC (Intergovernmental Panel on Climate Change) (2007), *Climate Change 2007 Synthesis Report: An Assessment of the Intergovernmental Panel on Climate Change*, Valencia, Spain, 17 November.

OECD (2001), *Assessing Microbial Safety of Drinking Waters: Perspectives for Improved Approaches and Methods*, OECD, Paris.

OECD (2003a), *Financing Strategies for Water and Environmental Infrastructure*, OECD, Paris.

OECD (2003b), *Improving Water Management: Recent OECD Experience*, OECD, Paris.

OECD (2003c), *Social Issues in the Provision and Pricing of Water Services*, OECD, Paris.

OECD (2006a), "Improving Water Management: Recent OECD Experience", *Policy Brief*, OECD, Paris.

OECD (2006b), *Infrastructure to 2030: Telecom, Land Transport, Water and Electricity*, OECD, Paris.

OECD (2006c), Revised Environmental Baseline for the OECD *Environmental Outlook to 2030*, Working Party on Global and Structural Policies, 24 October, ENV/EPOC/GSP(2006)23.

OECD (2006d), *Metrics for Assessing the Economic Benefits of Climate Change Policies: Sea Level Rise*, Working Party on Global and Structural Policies, ENV/EPOC/GSP(2006)3/FINAL, 26 July.

OECD (2008), *OECD Environmental Outlook to 2003*, OECD, Paris.

Prüss-Üstün, Annette *et al.* (2008), *Safer water, better health*, World Health Organisation, Geneva.

Siebert, Stefan *et al.* (2005), "Development and Validation of the Global Map of Irrigation Areas", *Hydrology and Earth System Sciences*, 9, 16 November, pp. 535–547.

Stern, Nicholas (2007), *The Economics of Climate Change: The Stern Review*, Cambridge University Press, Cambridge.

UNEP/GRID-Arendal Maps and Graphics Library (2006), *Global Costs of Extreme Weather Events*, accessed 13 November.

UNESCO (2006), Website with an extensive list of water-related events: *www.unesco.org/water/water_events/*, accessed in December.

United Nations (2006), *Water for All*.

UNSGAB (United Nations' Advisory Board on Water and Sanitation) (2006), accessed 3 May.

US Natural Resources Conservation Service (2006), National Water and Climate Center Website, accessed 2 May.

Warren, Rachel (2006), "Impacts of Global Climate Change at Different Annual Mean Global Temperature Increases" in Hans Joachim Schellnhuber (ed.), *Avoiding Dangerous Climate Change*, Cambridge University Press, pp. 93-131.

World Bank (2003), *Investing in infrastructure, what is needed from 2000 to 2030?*, Washington DC, July.

43

ISBN 978-92-64-05413-4
Space Technologies and Climate Change
Implications for Water Management, Marine Resources
and Maritime Transport
© OECD 2008

Chapter 3

Marine Resources and Maritime Transport: Trends and Outlook

Marine resources management and maritime transport both have strong links with climate change. Both activities can, on their own, lead to adverse environmental effects (i.e. pollution, depletion of marine resources with impacts on ecosystems), but both can be affected by climate change (i.e. extreme weather events, the opening of new sea routes near the poles). This chapter is divided into three main sections. A first introduces the intertwined marine and shipping activities with reference to the law of the sea; a second section focuses on marine resources management, particularly fishing and offshore activities; and a third provides information on maritime transport.

Introduction to marine resources and maritime transport

Marine resources management and maritime transport are two activities strongly linked in terms of their socio-economic impacts (Figure 3.1).[1] More than 90% of world trade involves transport via seas and oceans, while around 1 billion people (one-sixth of the world's population) depend on fishing activities (UNEP, 2007b).

Figure 3.1. **Overview of marine and maritime activities**

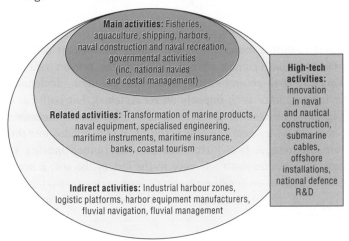

Source: Adapted from Kalaydjian, 2006.

Coastal countries have historically attributed significant value to "their" seas. According to the United Nations Convention on the Law of the Sea, a state's territorial sea extends up to 12 nautical miles (22 kilometres) including the airspace over it, its seabed and subsoil (UN Division for Ocean Affairs and the Law of the Sea, 2008).[2] Foreign vessels are allowed innocent passage through those waters. In addition, coastal countries can exercise their sovereignty over their exclusive economic zones (EEZ). The desire of coastal states to control the fish harvest in adjacent waters was a major driving force behind the creation of the EEZs. According to the Convention, the zone extends 200 nautical miles (370 kilometres) into the sea from the coastal baseline (see baseline in Figure 3.2) and forms a territory over which the coastal state has special exploration and marine resource rights. All other states have freedom of navigation and overflight in the EEZ, as well as freedom to lay submarine cables and pipelines.[3]

Figure 3.2. **Zones of national jurisdiction (Law of the Sea Convention, 1982)**
(nm = nautical miles)

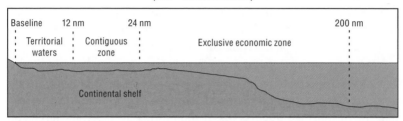

Almost all coastal countries of the world have delimited their exclusive economic zones, although in some cases geography is preventing delimitation, or disputes over specific EEZ boundaries are ongoing (*e.g.* the special case of the Mediterranean Sea).[4] Among the major beneficiaries of the EEZ regime are the United States, France, Indonesia, New Zealand, Australia and the Russian Federation (Table 3.1). The European Union members share their EEZ via the Common Fisheries Policy (CFP); thus vessels from one country can fish in another country's EEZ (with restrictions in territorial waters and protected zones), but must adhere to specific quotas on the amounts and types of fish they catch. The combined zone amounts to 25 million square kilometres, making it larger than those of any single country (EC, 2008a).

Table 3.1. **List of countries with exclusive economic zones (EEZ) by area size**

1	United States 11 351 000 km^2
2	France 11 035 000 km^2
3	Australia 8 148 250 km^2
4	Russian Federation 7 566 673 km^2
5	Japan 4 479 358 km^2
6	New Zealand 4 083 744 km^2
7	United Kingdom 3 973 760 km^2
8	Brazil 3 660 955 km^2
9	Canada 2 755 564 km^2
10	India 1 641 514 km^2
11	Argentina 1 159 063 km^2
12	Madagascar 1 225 259 km^2
13	China 877 019 km^2

In addition to exclusive economic zones, coastal States can under the UN Convention extend their sovereign rights even further over the continental shelf to explore and exploit the seabed up to 350 nautical miles (648 kilometres) from the coast, if this area extends beyond the EEZ. Countries need to follow a rather complex administrative and technical process set out in the Convention to demonstrate that their continental shelf extends beyond their established 200 nautical miles. Australia, France, Ireland, Mexico, New Zealand, Norway, the Russian Federation, Spain and the United Kingdom are among the countries that have launched the process. The United Nations special commission in charge of the procedure expects a high volume of submissions in the coming years (UN Commission on the Limits of the Continental Shelf, 2008).

Finally, more than a dozen regional fisheries management organisations (RFMOs) have been created to manage and conserve fish resources of the open seas. Governments that are members can share information and agree on management tools (i.e. quotas, vessel monitoring systems) for large zones outside their direct jurisdiction.

Many of the states that have established their EEZs or that are participating with RFMOs are not, however, in a position to exercise all their rights and perform duties under the Convention (e.g. prevent and limit pollution, facilitate marine scientific research). The delimitation of the EEZ, the surveying of its area, its monitoring, the utilisation of its resources and, more generally, its management and development are often seen as long-term endeavours beyond the capabilities of many countries, especially developing ones. But technical advances and new practices, particularly with the growing use of satellite technologies, are changing the situation for some countries, as will be shown in Chapter 4.

Despite technical and financial challenges, monitoring and surveillance of territorial waters and of EEZs have undeniably become crucial in recent years for several countries, particularly in light of the exploitation of natural resources. As marine and seabed resources become increasingly strategic, countries will implement more monitoring and defensive measures to protect those zones.

Marine resources

Marine resources management can be affected – as well as directly affect – climate change (Table 3.2). As mentioned by Ducrotoy and Elliott (2008), the main goal in managing the seas is to allow the sustainable use of its resources by human societies. But marine resources are by nature very diverse and unevenly exploited. They can be found at three different levels in the seas and oceans: the overlying waters, home to numerous biological resources (e.g. fish stocks); the top layers of the seabed, with polymetallic and phosphorite nodules (e.g. iron, nickel, copper, phosphates); and the underground seabed, with minerals – particularly hydrocarbons (e.g. petroleum). The majority of animal, vegetal and mineral resources can be found near the coastlines, as 75% to 95% of the animal and vegetal world exists in a ribbon no larger than 350 kilometres from the coasts, and at depths of less than 200 metres (French

Table 3.2. **Selected climate change connections with both fisheries management and exploitation of mineral resources in sea-floor**

	Possible climate change-induced effects on the activities	Possible effects of the activities on climate change
Fisheries management	• Extreme weather (causing damages to infrastructures and ships) • Introduction of new invasive species (jellyfish, algal bloom) via warmer waters • Acidification of waters in some areas, causing the displacement/ distribution of fish populations	• Endangering of fragile ecosystems through: – Overfishing – Depleting biological and vegetal marine resources, through invasive fishing techniques (*e.g.* destroying coral reefs) – Pollution (*e.g.* fishing vessels)
Exploitation of mineral resources in seafloor	• Extreme weather threatening oil rig installations and transport • New areas to dig for oil and gas (little/no more ice)	• Endangering of fragile ecosystems through: – Contributing to changes via new infrastructures (*e.g.* platform building) – Pollution (from offshore installations, new shipping routes to serve oil rigs, oil and gas accidents)

Academy of Sciences, 2003). This natural richness is endangered in many parts of the world by human activities, pollution and climate change-induced effects. The situation of fish stocks and minerals in the seabed in particular will be explored further in the next two sections.

Focus on fisheries

As stated earlier, around 1 billion people (one-sixth of the world's population) depend on fishing activities for their livelihood; in 2002 over 2.6 billion people received at least 20% of their animal protein primarily from marine sources (Hesse, 2005). Major changes in the fishing value chain have occurred around the world, due to globalisation, the multiplication of bilateral and multilateral agreements concerning EEZ and fishing stock management, fewer natural resources, pollution, and fleets' technological developments (OECD, 2008b). Certain space technologies have contributed to those trends, as will be explored later. One key message, however, is that for the fishing sector to remain sustainable on the regional and global scales, a number of measures are needed.

a) State of the world fisheries

There are no comprehensive data on the global fishing fleet size, but data from the FAO, OECD and other organisations indicate that the figure has increased rapidly between the 1950s and the early 2000s. Fleets have extended their operating range and adopted new technologies, such as advanced electronic fish finders.

Based on existing data, in 2004 the fleet consisted of about 4 million fishing craft; of these 1.3 million were decked vessels of various types, tonnage and power, and 2.7 million were open boats with no deck, including traditional craft operated by sail and oars (FAO, 2007). The data often exclude smaller boats whose registration is not compulsory and/or whose fishing licences are granted by local authorities. About 86% of the decked vessels were concentrated in Asia, followed by Europe (7.8%), North and Central America (3.8%), Africa (1.3%), South America (0.6%) and Oceania (0.4 %). Fishing communities are in general extremely vulnerable to economic shocks and changes in the market and supply. According to the FAO (2007), of the estimated 29 million fishermen worldwide, 5.8 million (20%) earn less than USD 1 a day. Their economic wellbeing therefore greatly depends on fishing abilities, marine resources and adequate weather conditions.

With regard to natural disasters, these events in themselves have led to short-term downturns for the commercial fishing sector. The resulting damage to equipment and infrastructure, however, can compromise the productivity of fishing and aquaculture activities. The 2005 tsunami in the Indian Ocean destroyed fishing boats, aquaculture pens and equipment. Developing countries suffer disproportionately from extreme weather events, as they often have weak response capacities. In 2000 the El Niño disaster reduced the amount of fish to 77-78 million tonnes (excluding China), the same level as in the early 1990s (FAO, 2002).

In terms of fish production, the world total was estimated at 140.5 million in 2004, with growth at 2.6% annually; the average consumption per person doubled since 1960 at the value of 16.2 kilograms per year (OECD, 2008a). This production includes marine capture fisheries, inland fresh water fisheries and aquaculture, all on the rise since the late 1980s. Concerning trade in fish products, the 2000 record of USD 55.2 billion was again topped in 2005 (OECD, 2005). Trade has steadily grown at a rate of 4% per annum in the last decade. China is the world's largest producer of fish products, as shown in Figure 3.3.

Four major FAO fishing areas produce almost 68% of the world marine catch (Figure 3.4). The northwest Pacific is the most productive, with a total catch of 21.6 million tonnes (25% of the total marine catch) in 2004; it is followed by the southeast Pacific, with a total catch of 15.4 million tonnes (18% of the marine total), and the western central Pacific and northeast Atlantic, with 11 million and 9.9 million tonnes (13% and 12%) respectively in the same year.

b) Worrisome trends for fishing activities

The three greatest threats to sustainability of fisheries resources and to the fishing sector are overfishing; illegal, unregulated, unreported fishing (IUU); and climate change impacts. If current trends continue, the world will increasingly face a scarcity of biological marine resources, with the additional challenge that climate change-induced impacts may be amplified.

Overfishing – Overfishing occurs when the number of fish catches in a year exceeds the minimum number of fish necessary for sustainable stock

Figure 3.3. **Marine and inland capture fisheries: top ten producer countries in 2004**

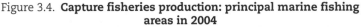

Norway 2.5
Thailand 2.8
Russian Federation 2.9
India 3.6
Japan 4.4
Indonesia 4.8
Chile 4.9
United States 5
Peru 9.6
China 16.9

Million Tonnes

Source: FAO, 2007.

Figure 3.4. **Capture fisheries production: principal marine fishing areas in 2004**

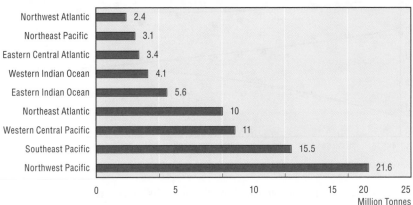

Northwest Atlantic 2.4
Northeast Pacific 3.1
Eastern Central Atlantic 3.4
Western Indian Ocean 4.1
Eastern Indian Ocean 5.6
Northeast Atlantic 10
Western Central Pacific 11
Southeast Pacific 15.5
Northwest Pacific 21.6

Million Tonnes

Note: Fishing areas listed are those with a production quantity equal to or more than 2 million tonnes in 2004.
Source: FAO, 2007.

development, hence causing imbalances between species composition in lakes, seas and oceans. By 2030, based on current exploitation trends (Figures 3.5 and 3.6), global individual consumption of fish and marine resources could surpass the current 16.2 kilograms per person, reaching an estimated 20 kilograms (Ifremer, 2007). However, due to the heavy concentration of consumption of a limited number of fish species, the top ten species that make up 30% of total fish consumed are already either fully exploited or overexploited (Hesse, 2005).

Figure 3.5. **Status of world fish stocks in 2005**

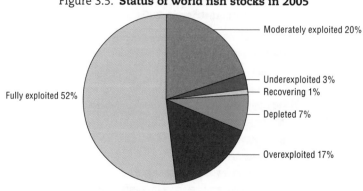

Moderately exploited 20%

Underexploited 3%
Recovering 1%

Fully exploited 52%

Depleted 7%

Overexploited 17%

Source: FAO, 2007.

Figure 3.6. **Global trends in the state of world marine stocks (1974-2006)**

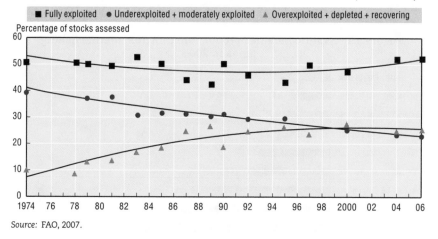

Source: FAO, 2007.

One tool to prevent overfishing is to establish sustainable management practices such as annual catch limits (*i.e.* the amount of fish allowed to be caught in a year). But this not enough. Countries need to make available the resources for developing reliable data on fish stocks and enforce their fishing regulations (OECD, 2008a). Many countries have developed fishing quotas based on scientific assessments of existing fish stocks for the maritime zones under their control, but control is often difficult over large areas and they must contend with pressures from fishing groups.

Illegal, unregulated, unreported fishing – IUU fishing activities constitute a growing worldwide problem that affects both domestic waters and the high seas, and undermines efforts to maintain sustainable fisheries worldwide (OECD, 2008a). The economics of those activities suggest they provide rather high profits to those who perpetrate them. As fish stocks become scarcer, partly because of

IUU fishing, fish quotas tend to decline further for law-abiding vessels, creating further incentives for legal operators to resort to IUU fishing (OECD, 2006). It is difficult to quantify the scale of the problem, as perpetrators can range from licensed fishers to large-scale organised crime organisations. There are several layers of IUU activities, responsibilities and possible responses:

- Illegal fishing activity is understood as fishing without a licence or fishing by contravening the terms and conditions of a licence (*e.g.* using illegal gear, catching over the allocated quota of fish, fishing in closed areas and/or seasons, exceeding by-catch limits). It is the coastal state's responsibility to deter illegal fishing through surveillance, detection, apprehension and the imposition of legal sanctions within its maritime areas (national waters), and to support the actions of its regional fishing organisations.

- Unreported fishing activities can also include a diversity of actors who misreport their fishing to the relevant national authority or regional organisation.

- Unregulated fishing includes fishing on the high seas by "free riders", *i.e.* those who fail to sign up to regional management arrangements and refuse to comply with the measures established by those arrangements. It also includes fishing on the high seas where there are no regional management arrangements in place (Marine Resources Assessment Group Ltd., 2005).

Figure 3.7. **Different types of illegal, unregulated, unreported (IUU) fishing**

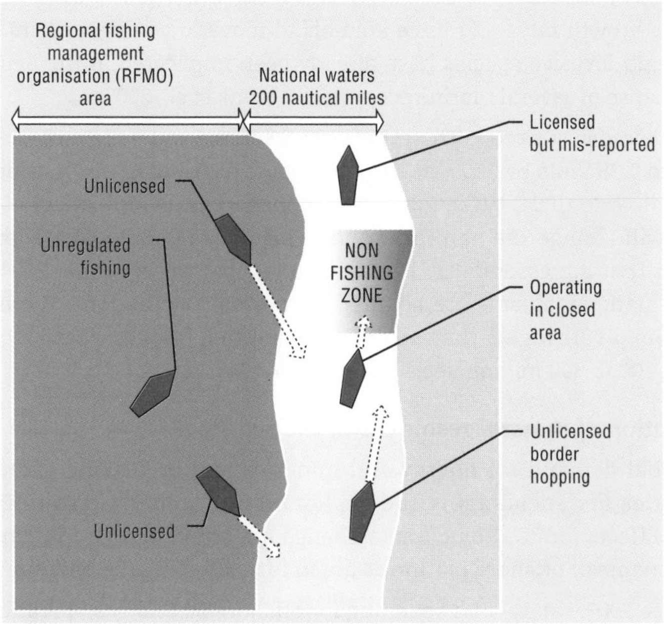

Source: Adapted from Marine Resources Assessment Group Ltd., 2005.

Climate change and fisheries – While there are regional differences, climate change will increasingly contribute to the transformation and loss of fishing stocks, notably via changes in sea temperature, acidification of waters and increasing extreme weather. The variability in environmental conditions and processes can have a strong influence on biological systems, which in turn may affect the marine ecosystem (leading to, *e.g.,* eutrophication) (Ducrotoy and Elliott, 2008).

Sea temperature: As the result of anthropogenic climate change, the mean temperature of sea surface waters is expected to increase and mean sea level is expected to rise (IPCC WG2, 2007). Based on current model simulations, there will very likely be a slowdown in the oceans' thermo-haline circulation by 2100, with severe consequences for fisheries and aquatic ecosystems. Already, the distributions of both exploited and nonexploited North Sea fishes have responded markedly to recent increases in sea temperature (Perry *et al.*, 2005). As ocean circulation drives larval transport, the recruitment patterns and population dynamics of marine organisms will be altered worldwide. Warming ocean temperatures influence species composition, breeding and population dynamics of plankton, benthos, fish and other species (UNEP GIWA, 2006). For example, the northeast Atlantic is experiencing an increased inflow of warm water that carries nutrition for cod and herring. These fish may thus increase in numbers as a result of greater nutrition. Nonetheless, since herring prey on other fish such as capelin, an increase in the herring population may decrease the capelin population in the long run. On the other hand, cold water may influence growth rates and force animals to move to warmer waters. Climate change and invasive species have already been implicated in the decline and even collapse of several marine ecosystems (Frank *et al.*, 2005).

Acidification of waters: The pH of ocean surface waters is projected to fall by 0.14 to 0.35 units by 2100, due to the uptake of rising levels of atmospheric carbon dioxide (IPCC WG2, 2007). The consequent acidification of surface waters will change the saturation horizons of aragonite, calcite and other minerals that are essential to calcifying organisms (Feely *et al.*, 2004). While many aquatic organisms are adapted to thermal fluctuations, the expected changes in pH are higher than any of those inferred from the fossil record over the past 200 to 300 million years (Caldeira and Wickett, 2003).

Exploitation of mineral resources in the sea-floor

Global demand for energy and minerals and increasing geotechnical capabilities are encouraging the exploitation of mineral resources in the seabed. Effects induced by climate change are affecting this industry, while developments of offshore platforms are in turn affecting the environment.

Since 2003, capital and spending for offshore oil and gas operations have grown at unprecedented rates, averaging 15% to 20% per year (Smith, 2008).

Figure 3.8. **Temperature increases and likely impacts on marine and terrestrial ecosystems**

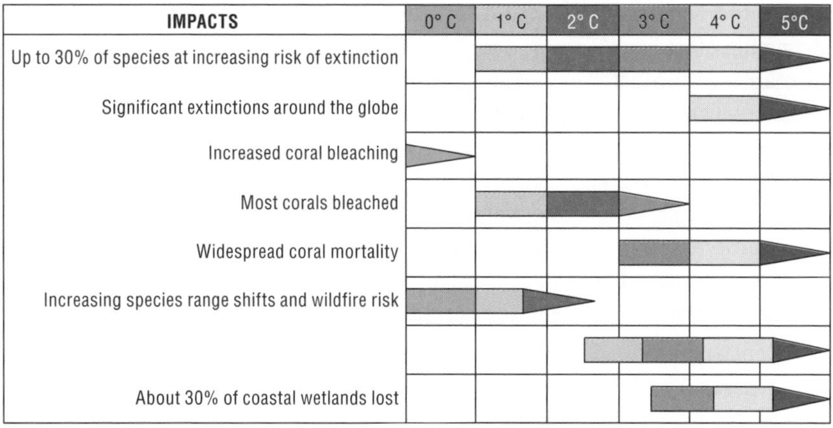

IMPACTS	0° C	1° C	2° C	3° C	4° C	5°C
Up to 30% of species at increasing risk of extinction						
Significant extinctions around the globe						
Increased coral bleaching						
Most corals bleached						
Widespread coral mortality						
Increasing species range shifts and wildfire risk						
About 30% of coastal wetlands lost						

Source: Derived from IPCC WG2, 2007.

This rise is due to increases in the price of energy and to the development of offshore drilling in deep-water environments: exploratory wells are costing on average around USD 40 million, with some reaching as high as USD 100 million in the most demanding regions. Already, 4 500 deep-water wells have been drilled globally since 1990. The first commercial offshore oil well, drilled by a mobile rig out of sight of land, took place in US territorial waters. The well was drilled in October 1947, in 14 feet (4.27 metres) of water in the Gulf of Mexico, off south eastern Louisiana (Gerwick, 2007). There are today over 6 500 oil and gas platforms in the Gulf of Mexico, and the US offshore oil and gas industry employs some 85 000 people.

In addition to the traditional actors in offshore drilling, more countries than ever before are developing their own production. The oil and gas industry is for instance the largest industry in Norway, followed by the maritime sector – and both sectors are still developing rapidly. As of March 2008, Norwegian oil rigs on order represented a value of some USD 10.5 billion for 33 rigs (Bøhler, 2008). To offer an example of a relative newcomer, thirty years ago Brazil was importing all its fuel for internal consumption. Today, using offshore drilling, that country is self-sufficient, and recent discoveries of a large reserve in the São Paulo bay (estimated at 33 billion barrels) could place Brazil among the top ten producers of oil. A technical challenge remains: the need to extract the oil below 2 000 metres of water and at least 4 000 metres of earth and sand.

One significant by-product of climate change is the rapid melting of polar ice caps, which had introduced the possibility of exploiting previously inaccessible territory and the seabed. The Arctic for instance is already becoming the object of competing claims among the five nations bordering that region, since the area could hold as much as one-quarter of the world's remaining undiscovered oil and gas deposits (Bird *et al.*, 2008). As seen in the introduction to this chapter, under the United Nations Convention on the Law of the Sea, a country claiming ownership of a region's ocean floor (generally with a view to developing offshore oil production) must show evidence that the seabed is an extension of their continental shelf. In that context, national scientific expeditions have multiplied in the Arctic region over the past five years (trying to demonstrate geographical claims). The Russian Federation and Denmark are, for example, looking particularly at the large underwater Lomonosov Ridge in the Arctic, to prove that large portions are part of their respective land masses (Borgerson, 2008). As a result of such seabed explorations for oil and gas, more scientific research is being conducted than ever before in the depths of the oceans.[5]

Construction of platforms at sea remains truly challenging due not only to the natural physical environment, but also to the requirements – despite increasing geotechnical capabilities – to dig further in the sea-floor (Gerwick, 2007). In addition, safe, efficient drilling operations depend on an accurate understanding of the current sea state and accurate and timely warning of impending storms. This growing industry is therefore dependent not only on real-time communications, but also on key meteorological forecasts. With regard to accidents, generally due to weather conditions or mechanical failure, there is a long-term trend towards increasingly severe average losses of mobile oil rigs since the early 1990s – even after discounting the loss impact of hurricanes Ivan, Katrina and Rita. But there was a significant increase in the overall insurance claims cost during 2004 and 2005, due mainly to an intense hurricane season in the Gulf of Mexico and strong storms in other regions (IUMI, 2008b).

Climate change-induced effects (*i.e.* extreme weather events, development of new sites for oil and gas production) could also feed on the long-term consequences of increased offshore drilling. The large-scale deterioration of the seas is a relatively recent phenomenon that has accelerated over the past fifty years. The causes of pollution are manifold – chemical, biological and bacterial – and represent an ecological and economic menace still not fully acknowledged. Warmer temperatures and sea rises in parts of the world may worsen the situation, while relatively new activities such as the exploitation of mineral resources in the seabed will also have impacts (Patin, 1999). Despite industry efforts such as banning the use of oil-based drilling fluids since 1992 (OSPAR Commission, 2008), the level of pollution from hydrocarbon spills and other chemicals from offshore oil and gas installations may well be underestimated, perpetuating the difficulty in assessing the scale of environmental effects (Fraser *et al.*, 2008).

*

*   *

Many scientific data remain to be collected and analysed, in order to understand better climate change's connections with the management of marine resources. As shown in this section, both fisheries management and exploitation of mineral resources in the sea-floor will be increasingly affected by, and may even themselves contribute to the impacts of climate change. One key lesson learned is the need to increase evidence-based knowledge about those specific sectors, particularly through collecting climate and environmental data and monitoring marine zones. The next section will focus on the growing maritime transport sector.

Maritime transport

The seas provide global transportation routes that link ports worldwide and are the lifelines of coastal economies and inland cities. As with marine activities, the links between maritime transport and climate change are sometimes subtle but strong. Maritime transport may both suffer (*e.g.* intensification of extreme weather events) and benefit (*e.g.* the opening of new sea routes) from climate change effects (Table 3.3). Maritime traffic is due to increase over the next two decades, and it will become increasingly important to monitor fleets and the effects of those fleets on ecosystems if seas and oceans are to continue sustaining human activities.

Table 3.3. **Climate change connections with maritime transport**

	Possible climate change-induced effects on the activities	Possible effects of the activities on climate change
Maritime transport	• Extreme weather events • Changes in currents • Opening of new sea routes	• Effects on ecosystems • Physical damages (especially to coral reefs) caused by running aground and anchoring • Introduction of invasive species • Marine pollution such as oil spills • Overcrowded shipping lanes • Increased air pollution

Source: OECD/IFP.

Continually growing traffic

The main driver for maritime transport growth is world trade. Maritime trade accounts for approximately two-thirds of merchandise trade (Grossmann *et al.*, 2007). For the past decade, strong world economic growth at an average of 4% has significantly contributed to the growth of the maritime trade. In 2006, GDP grew an average of 3% in developed countries, 6.9% in developing countries and 7.5% in transition economies (UNCTAD, 2007). At the same time, world merchandise trade grew by 8%, highlighting the effect of increasing globalisation and deepening economic integration. This high economic growth rate in 2006 has been encouraged and reinforced by transition economies and developing countries (mainly in Asia) that have been supplying primary commodities and raw materials, and that are emerging as important manufacturing centres. Maritime trade has been primarily driven by the growth in manufactured goods, which reached 72% of the total value of world exports in 2005 (USD 7.3 trillion out of a total of USD 10.1 trillion). Figures 3.9 and 3.10 highlight the positive correlation between world economic growth – particularly growth in merchandise exports – and seaborne trade.

Figure 3.9. **Indices for world economic growth (GDP), OECD industrial production, world merchandise exports (volume) and seaborne trade (volume), 1994-2006**

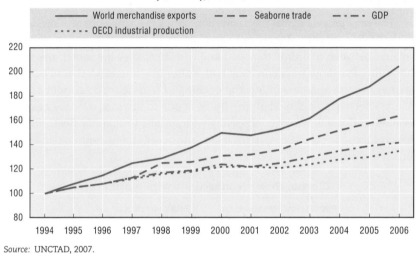

Source: UNCTAD, 2007.

Figure 3.10. **International seaborne trade for selected years**
Millions of tonnes loaded

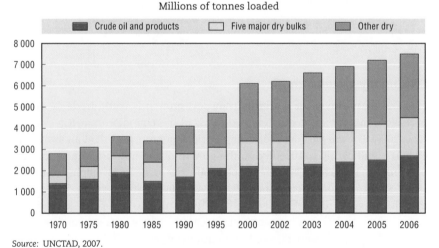

Source: UNCTAD, 2007.

SPACE TECHNOLOGIES AND CLIMATE CHANGE – ISBN 978-92-64-05413-4 – © OECD 2008

In 2006, the transport volume of seaborne world trade was estimated at 7.4 billion tonnes (compared to 6.78 billion tonnes in 2005) (UNCTAD, 2007). Crude oil accounted for 26.9% of total goods loaded, while petroleum products represented 9.2%. The larger balance of world goods loaded (63.9%) was made up of dry cargo, including bulk and containerised goods. A geographical breakdown of total goods loaded by continent highlights the continued predominance of Asia with a share of 39.1%, followed in descending order by America (21.5%), Europe (19.6%), Africa (10.7%) and Oceania (9.1%).

Both the number of ships and their capacities have continuously increased since the 1970s, in parallel with world economic growth. As Figure 3.11 graphically demonstrates, the number of ships has almost doubled and their capacity almost tripled between 1970 and 2004 (Grossmann et al., 2006). Most shipyards around the world are currently working to full capacity, and the world cargo fleet is set to grow by 243% by 2010 (Frank, 2008). The container fleet has reached 10 742 million TEU (twentyfoot equivalent units, a measurement of container's volume) in 2007, compared with only 3 766 million TEU a decade earlier in 1998. In comparison, scrapping of the tanker and bulker fleets remains very low, about 0.5% of the world fleet for both sectors (IUMI, 2008b).

Figure 3.11. **Development of the world merchant fleet**

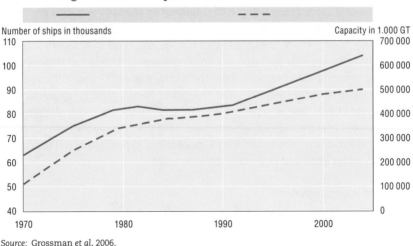

Source: Grossman et al, 2006.

Certainly, geographical advantages contributed to the growth of maritime transport in particular regions. As the European Union has access to four seas, its main mode of transport for foreign trade has been maritime. As a share of total transportation means for external EU trade, the share of sea trade has been dominant in both euro value and volume in tonnes (Table 3.4).

Table 3.4. **EU external foreign trade by mode of transport, 2004**

Mode of transport	Value in euros	Share in %	Volume in tonnes	Share in %
Sea	859.1	47.1	1 430	71.7
Road	259.7	14.2	100.8	5.1
Rail	25.1	1.4	89.3	4.5
Inland waterway	6.5	0.4	24.9	1.3
Pipeline	53.4	2.9	279.1	14.0
Air	471.7	26.0	9.8	0.5
Other	145.4	8.0	59.7	3.0
Total	1 822.9	100.0	1 993.6	100.0

Source: EC, 2005.

Looking ahead, maritime routes could become even busier. They are corridors of a few kilometres in width linking ports, and determined by a number of obligatory points of passage depending on physical constraints (marine currents, depth, reefs, ice) and political borders. As a result, these routes draw arcs on the earth's water surface. Seaborne trade could continue growing annually by 3.3% until 2020 (Grossman *et al.*, 2006), even if the growth of individual products within the maritime sector (*e.g.* liquid bulk *vs.* dry bulk) varies significantly due to the changing trade patterns of countries.[6]

Forecasts for the European Union alone suggest an increase of approximately 125% in the volume of shipment by sea between 2005 and 2030, a clear indication of the growing importance of maritime transport (Grossmann *et al.*, 2006). Today, 90% of Europe's external trade and close to 40% of its internal trade passes through its ports (EC, 2008b). Overall, global seaborne trade should continue growing more rapidly than the average growth rate, although less quickly than it will after 2015. On the other hand, greater expansion in intraregional trade flow is expected. Coastal and inland waterway shipping are also expected to show high expansion rates, as in many cases they are the focus of political subsidy programmes. This applies to the expansion of rail traffic too, albeit in milder form and regionally differentiated.

New sea routes opening

Sea route traffic is expected to see fundamental changes over the next decades, as climate change is opening new routes to access remote regions more easily.

Sea ice is an early indicator of climate change impacts. Sea ice covers around 10% of the world's oceans, and the measurements of its thickness, extent and composition help scientists understand and predict changes in weather and climate. A warmer earth has significantly contributed to the disappearance of the ice packs in the last decades. Sea ice could disappear in many parts of the world by 2050 if this trend continues.

In the Arctic for example, there are already well known recurring openings in sea ice (called *polynyas*), closing within hours or days. However based on the observed changes in the region, two important seasonal waterways circumventing the Artic, namely the Northwest Passage (near the United States' and Canada's coastlines) and the Northern Sea Route (near Norway and Russia's coastlines), are expected to become busier than ever (Figure 3.12). A larger use of the Northwest route for example is attractive, as it shortens transport distances between the Far East and European ports by 40%, allowing savings in journey times, fuel and carbon emission. With the disappearance of sea ice, the Arctic Ocean route would put Europe about 12 000 kilometres from the Far East, reducing for example by half the voyage time between Scandinavia and Japan. It would also cut sailing time from Germany to Alaska by 60%, going through Russia's Arctic instead of the Panama Canal.

Figure 3.12. **Current and future sea routes around the Arctic Basin**

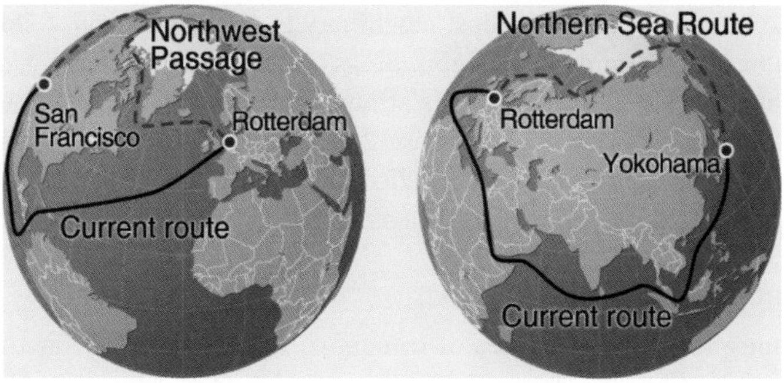

Source: UNEP, 2007c.

Weather constraints remain an important factor when analysing a sea route. In certain meteorological conditions, a ship will not be able to sail or will need to look for shelter while at sea. It then becomes necessary to know the potential time this type of ship would be out of service on those routes, where the wave height recorded exceeds the significant limit (Martinez et al., 2008). So ironically, even if new routes open at least partially thanks to climate change, induced extreme weather events may pose other difficulties for ships at sea, as shown in the next section.

Accidents at sea

Statistics on ship losses are not easily found; many accidents involving small vessels can go unreported, especially small-sized fishing and passenger-carrying ships. The more robust information comes from large commercial ship databases, and those show a rise in accidents, many linked to weather.

General cargo ships in particular run a high risk of accidents. They account for nearly 20% of the world merchant fleet, but suffer over 40% of total losses and almost 40% of the fatalities. These accidents amount to 90 ships lost and the death of 170 seafarers each year worldwide since 1995 (Royal Institution of Naval Architects, 2008). Although numbers of fatalities have been decreasing slightly since the early 2000s, there are still at times significant losses. For example, in the first three months of 2008 there were already more than 300 casualties worldwide (with loss of life for more than 100 seafarers and passengers); 88 of these casualties were caused by bad weather.

Weather remains the single greatest factor behind the total losses of ships, although collision and grounding are also common causes. Those losses have increased markedly in the past four years. The percentage of tonnage lost has almost doubled from 2005 (0.06%) to 2006 (0.11%), with 2007 at 0.08% (data available as of March 2008). However, those percentages still compare favourably with the 0.4% total losses of the early 1990s, thanks mainly to better communications equipment on board ships for emergency signalling. With regard to "partial" losses of ships, machinery is one key reason, followed by collision or contact damage and groundings, although bad weather is often an exacerbating factor. There has been a strong rise in the figures since 1998, with a 270% increase in one decade (from 247 serious incidents in 1998 to 914 in 2007). As a percentage of the world fleet increase, this is equivalent to 0.64% of the fleet suffering a serious partial loss in 1998 compared to 1.73% for 2007.

Pollution from sea traffic

Among the various modes of transport, ships remain by far the most energy-efficient and environmentally friendly means of conveying goods and passengers (Table 3.5). However, despite the enforcement of several international conventions (e.g. MARPOL, COLREG), some 20% of sea pollution still comes from the deliberate dumping of oil and other wastes from ships, from accidental spills, and from offshore oil drilling.

While accidental pollution at sea cannot be totally eliminated, operational routine pollution by ships is a common practice that causes major environmental problems in many parts of the world (Table 3.6). The rise in international maritime transportation and the lack of sea and coastal surveillance have led to increases in illegal discharges over the years. In addition, as international sea traffic has been rising rapidly for decades, ocean shipping pollution is also increasingly brought on by efforts to serve that sector through port infrastructure maintenance and fleet modernisation (OECD, 2008c). Another type of pollution, less known, comes from introducing invasive vegetal and biological species transported (i.e. in solid and water ballasts, and in hulls) and released by ships; this tips the sensitive species balance that often exists in a given ecosystem. The recent proliferation of

Table 3.5. **Energy use and CO_2 emissions associated with different modes of transport**

Mode of transport	Energy use (MegaJoule per tonne-kilometre)	CO_2 emissions (grams per tonne-kilometre)
Air	7-15	501-1073
Road	1.8-4.5	133-333
Rail	0.4-1	30-74
Inland barge	~0.3	~30-100
Sea[1]	0.1-0.4	3.5-7.7
oil products	0.1	7.7
dry bulk products	0.05	3.9
crude oil	0.045	3.5

1. Value reflects energy use and CO_2 emissions across a wide range of ships, including container ships. Data for non-container ships are estimates; actual energy use and emissions may differ from those shown.

Source: OECD, 2008b.

Table 3.6. **Episodic and routine pollution from ocean shipping activities**

Episodic environmental events	Routine environmental events
Vessel based:	
Oil spills	Engine air emission
Ocean dumping	Invasive species introductions (ballast water/hull fouling)
Sewage discharge	Toxic releases from hull coating
Oily wastewater	Underwater noise
Vessel collisions	
Impacts between ship and marine life	
Port based:	
Dredging	Stormwater runoff
Port expansion	Vessel wake erosion
Ship construction/breaking	Air emissions from cargo handling

Source: OECD, 2008c.

certain tropical algae in the northern hemisphere along main maritime routes, combined with warmer sea temperature, is an example of such pollution. As new sea routes open along the coastline of Siberia, it is expected that ships coming from warmer oceans will be carrying organisms that could become established in the Arctic Ocean (ACIA, 2005).

As mentioned by Saffache (2007), heavy sea traffic recorded in the English Channel, the North Sea (145 000 vessels per year) and the Mediterranean (8 000 vessels per year) accentuates the vulnerability of the coastline in terms of oil slicks. To limit pollution, a number of national and international regulations have been implemented to improve ships' design (*e.g.* double hulls to prevent oil spills, onboard treatment before discharge such as oily water separators) and sometimes new shipping rules (*e.g.* sea traffic separation schemes to avoid collisions). The costs of pollution remain high however, especially for episodic and large-scale accidental discharges. The 1989 Exxon Valdez oil tanker accident is still one of the largest oil spills and environmental disasters. Running aground, the tanker spilled 250 000 barrels into Alaskan water and onto shores. The spill cost has been assessed at around USD 7 billion, including the cleanup costs. More than USD 5 billion of this came from the largest punitive fine ever handed out to a company for corporate irresponsibility. But the social costs, including environmental, were deemed to be much larger, if not easily estimated.

Many hazardous or noxious substances are also carried in bulk form in tankers especially designed for that purpose. For example, half of all materials shipped through US waters are deemed hazardous. Ships carrying dangerous cargo are subject to Chapter VII of the International Convention for the Safety of Life at Sea, which regulates safety measures including such cargo's safe packaging and stowage. Identifying pollution from those ships is challenging, although tracking the ships during their voyage is one avenue pursued by many coastal authorities for prevention purposes. With a rising number of ships of all sizes, and the dependence of many countries on the inexorable growth in world seaborne trade, one key trend is the rise in ship values. Any sort of loss is assessed accordingly and the ship's value is reflected in the size of the ultimate claims.

Notes

1. Work on economic aspects of marine resources and maritime transport is currently being conducted in different divisions of the OECD, including the Trade and Agriculture Directorate and the Environment Directorate.

2. As of May 2008, 155 states have ratified the United Nations Convention on the Law of the Sea (1982). The Convention entered into force on 16 November 1994, and is today a globally recognised regime dealing with all matters relating to the law of the sea.

3. For example, the United States established a contiguous zone by a proclamation dated 2 September 1999,. This newly established zone, drawn according to international law and extending 24 nautical miles from the baselines of the United States, allows the country to exercise the control necessary to prevent infringement of its customs, fiscal, immigration and sanitation laws and regulations, as permitted in Article 33 of the Convention.

4. The Mediterranean, a semi-enclosed sea surrounded by 21 countries, has specific sea delimitation arrangements. Although most of the countries have declared territorial waters, few have claimed economic exclusive zones. As a result, the high seas area in the Mediterranean lies much closer to the coasts compared to most other seas and oceans on the planet.

5. In addition, many nations are currently working together to carry out exploration and monitoring of climate change in the Arctic and the Antarctic until 2009 for International Polar Year (e.g. via the Russian-American programme on Long term Census of the Arctic).

6. The HWWI forecasts are widely used to form predictive models for the future growth of maritime transport and trade. These models are based on past bilateral trade flows (1948-99) and the assumption that economic growth will contribute to an increase in trade relations among countries. Further assumptions include constant shares of cargo shipping by the various modes of transport as supported by historical observations (with the exception of liquid bulks which could be routed through pipelines).

Bibliography

Académie des Sciences (2003), *Exploitation et surexploitation des ressources marines vivantes*, Rapport sur la science et la technologie No. 17, Paris, France, December.

ACIA (Arctic Council and the International Arctic Science Committee) (2005), *Arctic Climate Impact Assessment studies*, Cambridge University Press, 2005.

Bird, Kenneth J., *et al.* (2008), "Circum-Arctic resource appraisal; estimates of undiscovered oil and gas north of the Arctic Circle: US Geological Survey Fact Sheet", 2008-3049, Version 1.0, July 23. *http://pubs.usgs.gov/fs/2008/3049/*, accessed July.

Bøhler, Karoline L. (2008), "Outlook on Norwegian Shipping", Norwegian Ship-owners' Association, IUMI meeting, Oslo, 10 March.

Borgerson, Scott G. (2008), "Arctic Meltdown: The Economic and Security Implications of Global Warming", *Foreign Affairs*, March/April.

Bryden, Harry L., Hannah R. Longworth and Stuart A. Cunningham (2005), "Slowing of the Atlantic Meridional Overturning Circulation at 25°-N", *Nature*, 438, pp. 655-657.

Caldeira, Ken and Michael E. Wickett (2003), "Oceanography: Anthropogenic Carbon and Ocean pH", *Nature*, 425, September.

Cicero, Anna Maria *et al.* (2003), "Monitoring of Environmental Impact Resulting from Offshore Oil and Gas Installations in the Adriatic Sea: Preliminary Evaluations", *Annali di chimica*, Vol. 93, No. 7-8, Società Chimica Italiana, Rome, Italy.

Ducrotoy, Jean-Paul and Michael Elliott (2008), "The Science and Management of the North Sea and the Baltic Sea: Natural History, Present Threats and Future Challenges", *Marine Pollution Bulletin*, Vol. 57, pp. 8-21.

EC (European Commission) (2005), *Energy and Transport in Figures 2005*, European Commission Directorate-General for Energy and Transport in co-operation with Eurostat, Brussels.

EC (2008a), *Common Fisheries Policy*, Website: *http://ec.europa.eu/fisheries/cfp_en.htm*, accessed 10 March.

EC (2008b), *An Integrated Maritime Policy for the European Union*, Communication from the Commission to the European Parliament, the European Council, the European Economic and Social Committee and the Committee of the Regions.

FAO (Food and Agriculture Organization) (2002), *La Situation mondiale des pêches et de l'aquaculture*, ftp://ftp.fao.org/docrep/fao/005/y7300f/y7300f01.pdf, Geneva, accessed 3 January.

FAO (2007), *The State of World Fisheries and Aquaculture 2006*, FAO, Rome.

Feely Richard A., *et al.* (2004), "Impact of Anthropogenic CO2 on the CaCO3 System in the Oceans", *Science*, 305(5682), pp. 362–366.

65

Firestone, J. and J.J. Corbett (2005), "Coastal and Port Environments: International Legal and Policy Responses to Reduce Ballast Water Introductions of Potentially Invasive Species", *Ocean Development and International Law*, 36(3), pp. 291-316.

Frank, Jerry (2008), "Insurers Fear Shipping Is Buckling Under Demand", *Lloyd's List*, 30 January.

Frank, K.T. *et al.* (2005), "Trophic Cascades in a Formerly Cod-dominated Ecosystem", *Science*, Vol. 308, pp. 1621-1623.

Franklin, Erik C. (2008), "An Assessment of Vessel Traffic Patterns in the Northwestern Hawaiian Islands between 1994 and 2004", *Marine Pollution Bulletin*, Vol. 56, pp. 136-162.

Fraser, G.S., J. Ellis and L. Hussain (2008), "An International Comparison of Governmental Disclosure of Hydrocarbon Spills from Offshore Oil and Gas Installations", *Marine Pollution Bulletin*, Vol. 56, pp. 9-13.

Gerwick, Ben C. (2007), *Construction of Marine and Offshore Structures*, Third Edition, March.

Grachev, Vladimir (2008), *Protection of the Environment in the Arctic Region*, Committee on the Environment, Agriculture and Local and Regional Affairs, Report 11477, Strasbourg, France, 3 January.

Grossmann, Harald *et al.* (2006), *Maritime Trade and Transport Logistics: Strategy 2030*, Hamburgisches Weltwirtschafts Institut.

Grossmann, Harald *et al.* (2007), "Growth Potential for Maritime Trade and Ports in Europe", *Intereconomics* , Vol. 227, July-August.

Häkkinen, Sirpa and Peter B. Rhines (2004), "Decline of Subpolar North Atlantic Circulation during the 1990s", *Science*, 304, pp. 555-559, 23 April.

Hesse, Stephen T. (2005), "Adapting to Sea Change: Managing Marine Resources in the Face of Climate Uncertainties", *Sustainable Development Law and Policy*, 37, Spring.

ICOADS (International Comprehensive Ocean–Atmosphere Data Set) (2008), Website *http://icoads.noaa.gov*, accessed 10 January.

Institut français de recherche pour l'exploitation de la mer (Ifremer) (2007), *The Fish Sector*, Website: *www.ifremer.fr/aquaculture/en/fish/index.htm*.

Institute of Shipping Economics and Logistic, *Shipping Statistics Yearbook*, ISL Universitätsallee, Bremen.

International Maritime Organization (2004), *International Convention for the Control and Management of Ships' Ballast Water and Sediments*, International Maritime Organization, London.

International Maritime Organization (2008), Website: *www.imo.org*.

International Maritime Organization and Marine Environment Protection Committee (2008), *Report of the Working Group on Annex VI and the NOx Technical Code*, edited by B. Wood-Thomas, London.

IPCC WG2 (Intergovernmental Panel on Climate Change Working Group II) (2007), *Working Group II Report: Impacts, Adaptation and Vulnerability*, Assessment of the Intergovernmental Panel on Climate Change, Valencia, Spain, 17 November.

IUMI (International Union of Marine Insurance) (2008a), "Dramatic Increase in Merchant Ship Total and Partial Losses", Press Release, Zurich, Switzerland, 19 March.

IUMI (2008b), "2007 Shipping Statistics Analysis: Total Losses Sharply Up and Major Partial Losses Continue to Rise", Report, Zurich, Switzerland.

Kalaydjian, Regis (2006), "L'économie maritime en Europe et en France" in Jean Guellec and Jean Lorot (eds.), *Planète océane : l'essentiel de la mer*, Editions Choiseul, Saint-Berthevin, France.

Kumar, S. and J. Hoffmann (2002), "Globalization: The Maritime Nexus", (Chapter 3) in C. Grammenos (ed.), *Handbook of Maritime Economics and Business*, Informa, Lloyds List Press, London, pp. 35-62.

Lindsey, Rebecca (2004), "A Dangerous Intersection: Scientists Recently Discovered a Link Between a Massive Coral Death in Indonesia, Man-Made Forest Fires, and El Niño", *NASA Earth Observatory*, April.

Marine Resources Assessment Group Ltd (2005), *Review of Impacts of Illegal, Unreported and Unregulated Fishing on Developing Countries, Final Report*, Prepared for the UK's Department for International Development (DFID), July.

McPhaden, Michael J. (2007), "Lessons Learned from El Niño and La Niña Monitoring", Presentation prepared for OECD Space Forum Working Group Meeting, 19 October.

Nicholls, R.J. *et al.* (2007), "Ranking Port Cities with High Exposure and Vulnerability to Climate Extremes Exposure Estimates", OECD Environment Directorate Working Papers, ENV/WKP(2007)1, 11 January.

Occhipinti-Ambrogi, Anna (2007), "Global Change and Marine Communities: Alien Species and Climate Change", *Marine Pollution Bulletin*, Vol. 55, pp. 342-352.

OECD (2005), *Review of Fisheries in OECD Countries: Policies and Summary Statistics*, 18, OECD, Paris.

OECD (2006), *Why Fish Piracy Persists: The Economics of Illegal, Unreported and Unregulated Fishing*, OECD, Paris.

OECD (2007), *The OECD Glossary of Statistical Terms*, OECD, Paris.

OECD (2008a), *OECD Environmental Outlook to 2030*, OECD, Paris.

OECD (2008b), "Transport in the Service of Trade and Climate Change: A Scoping Paper", Joint Working Party on Trade and Environment, OECD, Paris, June.

OECD (2008c), "The Impact of Globalisation on International Maritime Transport Activity: Past Trends and Future Perspectives", Working Party on National Environmental Policies, Working Group on Transport, ENV/EPOC/WPNEP/T(2008)2, OECD, Paris, April.

OSPAR Commission (2008), Website: *www.ospar.org*.

Overpeck, Jonathan T. *et al.* (2006), "Paleoclimatic Evidence for Future Ice-Sheet Instability and Rapid Sea-level Rise", *Science*, Vol. 24, March.

Patin, Stanislav (1999), *Environmental Impact of the Offshore Oil and Gas Industry*, EcoMonitor Publishing, New York.

Perry, A.L. *et al.* (2005), "Climate Change and Distribution Shifts in Marine Fishes", *Science*, Vol. 308, pp. 1912–1915.

Royal Institution of Naval Architects (2008), Website: *www.rina.org.uk*, accessed 5 March.

Saffache, Pascal (2007), "Reducing Marine Pollution: Fact or Fiction?", *Etudes Caribéennes*.

Schempf, F. Jay (2007), *Pioneering Offshore: The Early Years*, Pennwell Corp, September.

Selman, Mindy *et al.* (2008), *Eutrophication and Hypoxia in Coastal Areas: A Global Assessment of the State of Knowledge*, World Resources Institute, 1 March.

Smith, Michael (2008), "Escalating Offshore Expenditure, Production Expected", *OffShore*, Vol. 68, Issue 6, June.

Thomas, E.R., G.J. Marshall and J.R. McConnell (2008), "A Doubling in Snow Accumulation in the Western Antarctic Peninsula since 1850", *Geophysical Research Letters*, 35, L01706.

United Nations Commission on the Limits of the Continental Shelf (2008), Statement by the Chairman of the Commission on the Limits of the Continental Shelf on the Progress of Work in the Commission, Document CLCS/58, Twenty-first session, 17 March-18 April 2008, New York, 25 April.

UNCTAD (United Nations Conference on Trade and Development) (2007), *Review of Maritime Transport:Annual Report*, United Nations, New York.

UNEP (2007a), *Climate Change at a Glance*, Geneva, September, *www.unep.org/Themes/ climatechange/PDF/factsheets_English.pdf*.

UNEP (2007b), *Global Marine Assessments: A Survey of Global and Regional Assessments and Related Activities of the Marine Environment*, Geneva.

UNEP (2007c), *Global Outlook for Ice and Snow,* Geneva.

UNEP GIWA (United Nations Environmental Programme, Global International Waters Assessment) (2006), *International Waters: Regional Assessments in a Global Perspective*, February.

United Nations Division for Ocean Affairs and the Law of the Sea (2008), *Status of the United Nations Convention on the Law of the Sea*, Website: *www.un.org/Depts/los/ index.htm*, accessed 21 May.

Yin, Kedong *et al.* (1999), "Red Tides During Spring 1998 in Hong Kong: Is El Nino Responsible?", *Marine Ecology Progress Series*, Vol. 187, pp. 289-294, October.

SPACE TECHNOLOGIES AND CLIMATE CHANGE – ISBN 978-92-64-05413-4 – © OECD 2008

ISBN 978-92-64-05413-4
Space Technologies and Climate Change
Implications for Water Management, Marine Resources
and Maritime Transport
© OECD 2008

Chapter 4

Capabilities of Space Technologies

This chapter summarises the current contribution of space technologies to climate change research and monitoring, for fresh water, marine resources and maritime transport. Space systems and their ground infrastructure are tools that need to be used in combination with other assets. At the same time however, these systems have their own unique capabilities and can be put to uses ranging from snowmelt runoff measurement to improving safety at sea.

Introduction to satellite technologies

Space technology applications have begun to permeate many aspects of life in our modern societies. A growing number of activities – weather forecasting, global communications and broadcasting, disaster prevention and relief – increasingly depend on the unobtrusive utilisation of these technologies. Today over 30 countries have dedicated space programmes, and more than 50 have procured satellites. There are around 940 satellites operating in orbit; over two-thirds of them are communication satellites (OECD, 2007).

Launching a satellite into space to orbit either earth or another celestial body however remains a formidable challenge. Major progress has been achieved over the past few decades, including notably the successful development of several families of launchers (*e.g.* Soyuz, Ariane, Atlas, Delta), but access to space remains costly and risky. Satellites are basically platforms that can carry instruments used for diverse applications. They are often very sophisticated R&D objects with a lengthy development time (several years), although the greater recurring use of standard satellite platforms is reducing that time (six months or less for some small satellites).

Scientific research and space exploration remain two key objectives of satellite development, but the strategic and economic significance of many down-to-earth technical uses has grown, mainly because of the capacities to:

- Communicate anywhere in the world and disseminate information over broad areas, whatever the state of the ground-based network.
- Observe any spot on earth accurately and in a broad spectrum of frequencies, in a non-intrusive way.
- Locate with increasing levels of precision a fixed or moving object anywhere on the surface of the globe.

Space technologies therefore boast unique capabilities. However, there are a number of technical constraints that may lessen the usefulness of satellite signals or data for specific applications (See Annex B for basic information about the different types of satellite sensors). Primary among these limitations are the following:

- The geographic area that a sensor can cover in one satellite pass and the level of detail that can be seen (a function of the satellite swath width, orbit and sensor's resolution – as with a telescope, the more one zooms, the less global coverage one gets).

- The satellite's revisit time over one specific area (from many times a day to only once a month depending on the orbit chosen for the satellite – see Table 4.1).

- The adequacy of the onboard sensors for a particular element that needs to be observed (this depends on the choice of sensors carried on the satellite, optical or radar, and on the bands that figure in the electromagnetic spectrum).

Table 4.1. **Basics about satellites' orbits**

Orbit	Description
Low earth orbit (LEO)	Satellites in LEO orbit the earth at altitudes of between 200 km and 1 600 km. Compared with higher orbits, LEO satellites can capture images and data with better detail (better resolution), have speedier communications with earth (less latency), and require less power to transmit their data and signals to earth. However, due to friction with the atmosphere, a LEO satellite will lose speed and altitude more rapidly than in higher orbits.
Polar orbit	A majority of satellites never "see" the poles, as more often than not they are positioned in equatorial orbits to cover large populated areas. Satellites that use the polar orbit – particularly meteorological satellites – go over both the North and the South Pole at a 90-degree angle to the equator. Most polar orbits are in LEO, but any altitude can be used.
Geosynchronous/ Geostationary orbit (GSO/GEO)	The satellites in geosynchronous orbit (also known as geostationary when it has an inclination of zero degrees) are at a higher altitude, around 36 000 kilometres, forming a ring around the equator. Their orbits keep them synchronised with the earth's rotation, hence they appear to remain stationary over a fixed position on earth, and provide an almost hemispheric view. Their advantage is the frequency with which they can monitor events (three GEO satellites placed equidistantly can together view the entire earth surface, but with less precision than LEO satellites). They are ideal for some types of communication and global meteorological coverage.
Sun Synchronous Orbit (SSO)	When in sun-synchronous orbit, the satellite orbital plane's rotation matches the rotation of the earth around the sun and passes over a point on earth at the same local solar time each day.

Note: Orbital mechanics (also called flight mechanics) deal with the motion of artificial satellites and space vehicles moving under the influence of forces such as gravity, atmospheric drag, thrust, etc. There are many types of orbits other than the ones described above [e.g. medium earth orbit (MEO) for some navigational and communications satellites, Molniya orbits, etc.].

Scientific knowledge about climate change, fresh water and the oceans

The need to gain scientific knowledge about climate change, fresh water and the oceans, as well as the interactions among these, is increasing as changing weather patterns are having diverse impacts around the world. Many activities also call for operational monitoring of the oceans. This section reviews the contributions of space technologies.

An unforeseen but now indispensable role for satellites

In the context of climate change, calls for sound data on the state of the environment from decision makers and the public are growing. Since the beginning of the space age, space-based observations of climate variables have been made by operational meteorological systems and satellite R&D missions. Satellite capabilities – whether for dedicated missions or not – are now increasingly matching data requirements.

The scientific accomplishments enabled by satellite use are numerous. Reliance on space-based observations, telecommunications and navigation has grown internationally over the years, although significant gaps remain in data measurement capabilities and their continuity. For many years, R&D satellite missions were not specifically designed for climate observation and monitoring, and some scientific breakthroughs have been almost accidental. Two examples provided below concern findings on the earth's ozone layer and global precipitation measurements.

In the early days of civilian meteorological satellites in the late 1970s, a NASA mission was launched to study weather patterns by mapping global ozone (Spector, 2007). The Nimbus 7 satellite carried a new R&D sensor dubbed the Total Ozone Mapping Spectrometer (TOMS). Scientists soon realised that some of the data collected by TOMS were more significant than they had initially anticipated, as the instrument allowed them to study ozone in the upper and lower atmosphere in a way that had never been done before, more frequently, and with far greater detail. The data from TOMS led to the detection of long-term damage to the ozone layer above the Arctic, the Antarctic and heavily populated areas. Research conducted using the data led to the passage of the Montreal Protocol in 1987, an international agreement restricting the production of ozone-depleting chemicals. Originally intended to be operational just a few years, the TOMS instrument was not retired until 2007 after more than thirty years of use on different satellites. It has been succeeded by the Ozone Monitoring Instrument, a more advanced spectrometer that currently flies on NASA's Aura satellite.

A second illustration is the Tropical Rainfall Measuring Mission (TRMM) satellite, another recent "example of unexpected 'bonuses' often accruing from a scientific mission" (NRC, 2008). The TRMM satellite was launched in 1997 by NASA in co-operation with the Japan Aerospace Exploration Agency. The mission demonstrated the feasibility of obtaining near-real-time global coverage of precipitation observations from space. This made meteorologists better able to measure spatial and temporal variability of rainfall in the tropics, especially over the oceans (NRC, 2008). TRMM also proved the worth of a number of technologies (*e.g.* the feasibility of paired radar and passive microwave systems in space) and yielded scientific results (*e.g.* showing that a multisensor reference satellite could calibrate data from other space-based observational systems; mapping by radar of the three-dimensional structure of precipitating weather systems and so predicting, *inter alia,* tropical cyclone tracks and intensity). The TRMM mission was originally intended to last three to five years but has enough fuel to remain in orbit until 2012. It could be followed by the Global Precipitation Measurement Mission in 2013. The GPM Mission includes a core satellite that makes measurements and carries dual-frequency precipitation radar and a passive microwave sensor. The data from this satellite are to be inter-calibrated with those from international operational and research satellites carrying similar microwave sensors. The overall system could provide global estimates of precipitation approximately every three hours.

There are many other examples of past and current scientific accomplishments where satellite data played a role (Table 4.2).

Table 4.2. **Examples of scientific accomplishments involving earth observations and landmark satellites**

Accomplishment	Satellites
Monitoring global stratospheric ozone depletion, including Antarctica and Arctic regions	TIROS series, Nimbus 4 and 7, ERS 1, ERS-2, Envisat
Detecting tropospheric ozone	Nimbus 7, ERS 2, Envisat, Aqua, Aura, MetOp
Measuring the earth's radiation budget	Explorer 7, TIROS, and Nimbus
Generating synoptic weather imagery	TIROS series, ATS, SMS, MetOp
Assimilating data for sophisticated numerical weather prediction	Numerous weather satellites, including the TIROS series, NOAA's GOES and POES, Eumetsat's MetOp, ERS 1, ERS-2, Envisat
Discovering the dynamics of ice sheet flows in Antarctica and Greenland	Radarsat, Landsat, Aura, Terra, Jason, ERS-1, ERS-2, Envisat
Detecting mesoscale variability of ocean surface topography and its importance in ocean mixing	Topex/Poseidon, ERS 1, ERS-2, Envisat
Observing the role of the ocean in climate variability	TIORS-N and NOAA series, ERS 1, ERS-2, Envisat
Monitoring agricultural lands (a contribution to the Famine Early Warning System)	Landsat, Spot series
Determining the earth reference frame with unprecedented accuracy	Lageos , GPS

Source: Adapted from NRC, 2008.

The main advantages of using satellites for climate change can be summarised (Payne *et al.*, 2006):

● Data are collected year-round and can provide information when field data collection is not possible, due to remote locations limiting on-the-ground access or to bad weather conditions.

● For certain applications there may be reduced costs when compared to traditional field data collection methods in remote environments (land cover classification for example).

● Given the geographic extent of oceans, *in situ* observations alone cannot adequately characterise dynamics. Remote sensing systems can capture a synoptic view of the landscape.

● Remote sensing provides additional information that can supplement more intensive sampling efforts and help extrapolate findings.

Responding to specific requirements for climate and ocean variables

This section reviews in more detail some of the current data required by scientists and decision makers with regard to climate and ocean variables, and determines whether they may rely on space-based systems. It appears that in many cases satellites have furnished very useful data over the years, although some inherent limitations still exist.

Essential climate variables

Existing satellites collect data on key earth parameters – or essential climate variables, ECVs –from the atmosphere, oceans, and land. A large number of those variables are used by national and international groups, such as the United Nations Framework Convention on Climate Change (UNFCCC), to research and monitor the climate. Over the years several international co-ordination bodies – in particular the Committee on Earth Observation Satellites (CEOS) and the Global Climate Observing System (GCOS) – have, together with space agencies and scientific groups, developed detailed requirements for satellite-based climate observations.[1]

One important aspect is that sensors all have their particular strengths and weaknesses. Thus no one sensor can be used for every aspect of monitoring, because of different spatial resolution (detailed metric vs. large kilometric view), spectral resolution (infrared vs. other bands), temporal resolution (revisit time), and atmospheric conditions. Even when tasked with the same mission, space systems usually have their respective specificities. For example, there are currently several satellite systems from different countries measuring sea surface temperature, but each instrument measures the temperature at a slightly different depth and with very different accuracy (ESA, 2008).

Despite those limitations, a large number of essential climate variables depend on satellite data. Based on a list submitted by the GCOS at the request of the UNFCCC to provide a global and internationally agreed set of climate change variables, a total of 34 ECVs benefit from space observations (Table 4.3).

Overview of the information flow in water remote sensing

Remote sensing from space relies on several parameters to draw data from water bodies. Many physical and biological phenomena have a "surface signature". This means that large lakes or snow cover for example provide signals that can be detected from space.

To generalise, a space-borne sensor carried onboard a satellite detects in its fields of view the surface signature of a water-related phenomenon and reacts to it (Figure 4.1). Despite inherent noise and disturbances in the atmosphere, the sensor receives the information and then transmits it to the

Table 4.3. **Thirty-four essential climate variables (ECVs)
and their dependence on satellite observations**

Those largely dependent in italic

Domain		Essential Climate Variables (ECV)
Atmospheric (over land, sea and ice)	**Surface:**	Air temperature, *precipitation*, air pressure, surface radiation budget, wind speed and direction, water vapour
	Upper-air:	*Earth radiation budget (including solar irradiance), upper-air temperature [including Microwave Sounding Unit (MSU) radiances], wind speed and direction (especially over the oceans), water vapour, cloud properties*
	Composition:	*Carbon dioxide*, methane, *ozone*, other lasting greenhouse gases, *aerosol properties*
Oceanic	**Surface:**	*Sea surface temperature*, sea surface salinity, *sea level, sea state, sea ice*, current, *ocean colour (for biological activity)*, carbon dioxide partial pressure
	Sub-surface:	Temperature, *salinity*, current, nutrients, carbon, ocean tracers, phytoplankton
Terrestrial		River discharge, water use, ground water, lake levels, *snow cover, glaciers and ice caps*, permafrost and seasonally frozen ground, *albedo, land cover (including vegetation type)*, fraction of absorbed photosynthetically active radiation (FAPAR), leaf area index (LAI), *biomass, fire disturbance, soil moisture*

Source: CEOS, 2006.

Figure 4.1. **Information flow in water remote sensing**

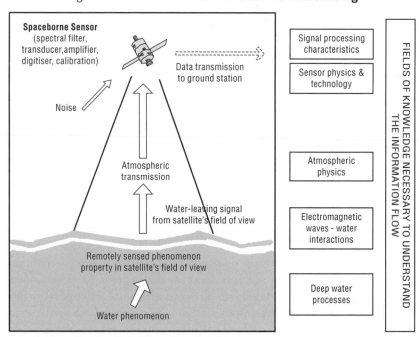

Source: Adapted from Robinson, 2004.

ground station. Only properties at the water surface can be detected, although in some cases observing the surface can also provide some information about what is happening beneath.

As an example, internal waves – a dynamic phenomenon centred dozens of metres below the sea surface – can be revealed in considerable spatial detail in certain radar imagery because of their surface roughness signature (Robinson, 2004). For the "water colour", which helps determine water quality and detect pollution, the uppermost few metres can be observed; the temperature, roughness and water height parameters can only be observed at the surface. To provide a proper interpretation of the observed surface signatures, a number of fields of knowledge presented in the right column of the figure (e.g. sensor physics and technology) need to be taken into consideration.

It is also possible to measure elements of the global water cycle using diverse space-based systems. The estimated residence times for water range from one week (e.g. biospheric water) to 10 000 years (e.g. ground water) – hence the need for reactive, timely and long-term observations. Among the key parameters of the overall water cycle, four are particularly important:

- *Precipitation* occurs in a variety of forms (i.e. rain, hail, freezing rain, sleet, snow).

- *Evaporation* is the physical process by which a liquid (i.e. water) or solid substance (i.e. snow) is transformed to the gaseous state.

- *Evapotranspiration* can be defined a bit differently depending on the scientific angle used, but it consists generally of the sum of transpiration (i.e. evaporation of water from plant leaves) and evaporation from water in the soil (e.g. rivers, lakes).

- *Water runoff* is composed of a mixture of water and soil (along with any other organic or inorganic substances); it is caused by precipitation, snowmelt, over-irrigation, or other water coming in contact with the earth and carrying matter to streams, rivers, lakes and other surface water bodies. A water runoff can be a source of pollution, carrying polluted substances over large areas in the case of storms.

Studying the atmosphere (cloud, water vapour, precipitation and wind)

Measuring the water in the atmosphere contributes to existing knowledge about the global water cycle and sends early warnings of possible problems, such as drought or floods due to lack of or excessive precipitation and pollution. Several space-borne systems are used to monitor clouds, water vapours, precipitation and winds on different spatial and temporal scales, although ground systems are still the main components of the architecture (e.g. networks of pluviometers for continuous direct local measurements, ground-based radars).

Cloud cover – Cloud cover information from space has been provided over the past decade mainly from three instruments flying on different US and European satellites: the Advanced Very High Resolution Radiometer (AVHRR), the Moderate Resolution Imaging Spectrometer (MODIS) and ESA's Medium Resolution Imaging Spectrometer (MERIS) (ISU, 2004). Those instruments were not primarily intended to measure clouds, so they can experience difficulties in areas of heavy cloud cover. As of June 2008, five different satellites – four from NASA and one from CNES dubbed the "Afternoon Constellation" or "A-Train" – provide data on the interactions of clouds, pollution and rainfall (Figure 4.2). The satellites, which each carry the AVHRR, MODIS or MERIS instruments, pass over the equator within a few minutes of each other in the afternoon around 13:30 (local time) each day, allowing for near-simultaneous observations and measurements of clouds over a precise area by different means. The combination of data from those satellites and others has already provided (in early 2008) some unforeseen findings on the effect of pollution on clouds. A NASA team showed that South American clouds infused with high levels of carbon monoxide due to power plants or agricultural fires – classified as "polluted clouds" – tend to produce less rain than their clean counterparts during the region's dry season (NASA Goddard, 2008). More studies of the interactions between aerosols and clouds are needed to understand pollution dynamics better, and this will be facilitated as more diverse data become available.

Figure 4.2. **The "A-Train" satellite constellation studying the atmosphere**

Note: As of June 2008, five satellites are in orbit: NASA's Aqua, Aura, CloudSat, CALIPSO and CNES' PARASOL.
Source: NASA Goddard, 2008.

Water vapour in the atmosphere – Water molecules from the ocean, soil and vegetation circulate in the atmosphere through different processes, such as evaporation. Studying water vapour not only adds to the understanding of climate processes but also contributes to weather forecasting. Ground-based probe measurements are still the routine way to observe the vertical distribution of water vapour. However, the procedures for deriving data from remote sensing (*i.e.* using remotely sensed thermal infrared measurements) and navigation satellite signals (a promising technique – see Box 4.1), and then integrating them in models, have gone from experimental to quasi-operational. Infrared sensors on satellites can measure water vapour in a layer of atmosphere several kilometres above the earth's surface. Data are being received from a number of radiometer sensors that are carried on board meteorological satellites, such as NOAA's polar satellites (*i.e.* Advanced Microwave Sounding Unit B on NOAA-15, NOAA-16, NOAA-17) and the US Department of Defense's Meteorological Satellite Programme satellites (*i.e.* Special Sensor Microwave Imager or SSM/I). Based on these data and new atmospheric models, new climate patterns called atmospheric rivers have been identified (Neiman *et al.*, 2008). These are long plumes of moisture streaming over large bodies of water, which can be responsible for either heavy winter rains, major floods, or – when they are almost absent – extreme episodes of dry, hot weather in summers.

Box 4.1. **Improving weather forecasting by measuring water vapour**

Launched in April 2006, the US-Chinese Taipei COSMIC constellation (Constellation Observing System for Meteorology, Ionosphere and Climate) is a system of six micro satellites; these measure the bending of radio signals from the US global positioning system (GPS) as the signals pass through the earth's atmosphere. COSMIC can see through cloud cover and gather highly accurate data through many levels of atmosphere. Initial results show that the system's unique global coverage provides unprecedented information on the atmosphere's temperature and water vapour structure. Moreover, COSMIC data can be collected above hard-to-reach locations, such as Antarctica and the remote Pacific; that could greatly enhance the global-scale monitoring needed to analyse climate change. According to scientists working with COSMIC data, after only a few months it is possible to see the strengths and weaknesses in some forecasting models that were not possible to see before.

Source: National Center for Atmospheric Research, 2006.

Precipitation – Concerning precipitation, the data provided by ground-based measurement systems (*e.g.* meteorological and radar stations) have proved cost-effective in continuously monitoring precipitation. However, it is not possible for these ground-based systems to monitor weather conditions in remote areas, over the oceans or in less developed regions of the world. The first satellite mission dedicated to precipitation studies was NASA-JAXA's Tropical Rain Monitoring Mission (TRMM), mentioned above. In June 2008, Precipitation Radar (PR) is still the only such radar in space, providing direct, fine-scale observations of the three-dimensional structure of precipitation systems (i.e. unique vertical profiles of the rain and snow from the surface up to a height of about 20 kilometres), with high revisit time.

Wind – Satellite ocean surface vector wind (OSVW) data have greatly influenced how winds are measured over the oceans, particularly with the improvement of operational weather forecasting and warning capabilities (e.g. for hurricanes) (Jelenak and Chang, 2008). The first wind-measuring microwave radar instrument in space (with data used operationally by meteorological offices) was the scatterometer launched on ESA's ERS-1 in 1991. It was followed by NASA's Scatterometer (NSCAT), launched in 1996 aboard the Advanced Earth Observing Satellite (ADEOS). It provided continuous measurements of ocean surface wind speeds, taking 190 000 wind measurements per day and mapping over 90% of the world's ice-free oceans every two days. In doing so, it furnished more than 100 times the amount of ocean wind information then available from traditional ships' reports (NASA JPL, 1996). This mission was followed in 1999 by the launch of QuikSCAT, which provided even more detailed ocean winds data. A QuikSCAT follow-on mission is currently being considered, so as to allow data continuity and improve OSVW measurement capabilities (Gaston and Rodriguez, 2008).

Soil moisture and salinity

Soil moisture and sea/ocean salinity are important variables for climate monitoring and for efficient water management practices in local and regional communities. These parameters also significantly affect the global balance in energy and moisture, hence providing important information for climate modelling. Currently there are relatively few datasets on either soil moisture or water salinity.

Soil moisture – This is a critical component of temperature and precipitation forecasts, as well as other applications. Soil moisture measurements are usually required to depths of 1 to 2 metres, an area often referred to as the "root zone". Soil moisture content affects infiltration (i.e. the process of water entry from surface sources such as rainfall or irrigation into the soil), runoff, and the water available for use by vegetation. Measurement of the moisture profile through the plant rooting zone and below is, therefore, of the utmost importance. However, satellites are not capable of performing direct measurements of the moisture throughout the soil profile below the thin surface layer. Space-based measurements provide estimates of water in the upper 5-10 centimetres of soil.

Table 4.4. **List of remotely sensed oceanographic parameters,
their observational class, and representative satellites
and sensors they carry**

Parameter	Observational category	Satellites (sensors)
Bio-optical properties of oceanic waters (*e.g.* ocean colour)	Visible – near infrared	Envisat (MERIS), AQUA (MODIS), Orbview2 (SeaWifs)
Bathymetry (*i.e.* measurement of the depth of the ocean floor from the water surface)	Visible – near infrared	Landsat, Spot, Ikonos
Sea surface temperature	Thermal infrared Microwave radiometers	POES (AVHRR), GOES (Imager), DMSP (SSM/I), TRMM (TMI), MetOp
Sea surface salinity	Microwave radiometers and scatterometers . .	
Sea surface roughness, wind velocities, waves and tides	Microwave scatterometers and altimeters Synthetic aperture radar	ERS-2, QuickSat, RADARSAT-1, Envisat
Sea surface height, wind speeds	Altimeters	Topex-Poseidon, Jason-1, QuikSCAT
Sea ice	Visible – near infrared Microwave radiometers, scatterometers and altimeters Synthetic aperture radar	POES (AVHRR), DMSP (SSM, I), ERS-1, ERS-2, RADARSAT-1 and 2, Envisat
Surface currents, fronts and circulation	Visible – near infrared	POES (AVHRR), GOES (Imager), Topex-Poseidon, Jason-1, Envisat
Surface objects, ships, wakes and flotsam	Synthetic aperture radar	RADARSAT-1, Envisat (ASAR)

Source: Adapted and updated from Brown *et al.*, 2005.

While these suffice for some applications, many others require moisture measurements through the soil profile (CEOS, 2006). As a result, current efforts are focusing on improving hydrological models to understand and, if possible, quantify the relationship between surface soil wetness and subterranean soil moisture. In 2009, the Soil Moisture and Ocean Salinity (SMOS) mission, the second Earth Explorer mission to be developed as part of ESA's Living Planet Programme, should allow observation of soil moisture over the earth's landmasses and salinity over the oceans. Since SMOS will provide data on soil moisture content to a depth of only a few centimetres, modelling techniques are being developed to derive the moisture content within the root zone from time series of near-surface soil moisture.[2]

Salinity – Salinity can be detected from space-borne sensors only when it occurs on the surface or in the root zone of vegetation, so satellites cannot map soil salinity directly. They are used to map other parameters – such as optical reflectance at a fixed wavelength, natural gamma radiation, and electric conductivity – from which salinity models can then be derived. Demonstrations are still under way using other already orbiting multispectral satellites (*e.g.* Landsat, Spot, IRS, Ikonos) to help detect exposed salt on the surface (*e.g.* monitoring the decline in photosynthetic activity in vegetation as a result of the salt stress; detecting salinity from the type of the vegetation, which can indicate the absence of salt in the root zone). The SMOS satellite,

Table 4.5. **Overview of selected space systems with ocean-related missions**

Mission	Dates	Description
Tropical Rainfall Measuring Mission (TRMM)	1997 – present (extended until 2009, could be operational until 2012)	TRMM is a joint mission between NASA and Japan (JAXA) with five instruments, including the first Precipitation Radar (PR) in space. These instruments are used to monitor and predict tropical cyclone tracks and intensity, estimate rainfall, and monitor climate variability (precipitation and sea surface temperature).
SeaWinds (on board QuickScat satellite)	1999 – present	The SeaWinds instrument on the QuikScat mission is a quick recovery mission to fill the gap created by the loss of data from the NASA Scatterometer (NSCAT) when the ADEOS satellite lost power in June 1997. The SeaWinds instrument is a specialised microwave radar that measures near-surface wind speed and direction under all weather and cloud conditions over the earth's oceans.
Jason-1 and -2	2001 – present	Jason is a joint United States-France (CNES) oceanography mission designed to monitor global ocean circulation, discover the tie between the oceans and atmosphere, improve global climate predictions, and monitor events such as El Niño and La Niña and ocean eddies. Its successor, Jason-2, was launched in June 2008.
SeaWiFS (onboard OrbView-2 satellite)	1997 – present	The purpose of the Sea-viewing Wide Field-of-view Sensor (SeaWiFS) project is to furnish quantitative data on global ocean bio-optical properties to the earth science community. Subtle changes in ocean colour signify various types and quantities of marine phytoplankton (microscopic marine plants); this knowledge has both scientific and practical applications. NASA purchase the OrbView-2 data from Orbimage.
Aqua	2002 – present	The Aqua satellite carries six instruments, including the Moderate Resolution Imaging Spectroradiometer (MODIS) also carried by the Terra satellite, and observes the earth's oceans, atmosphere, land, ice and snow covers, and vegetation.
SeaWinds (on board ADEOS II)	2002 – present	The Advanced Earth Observing Satellite II (ADEOS II), the successor to the Advanced Earth Observing Satellite (ADEOS) mission, is a NASA-JAXA joint mission. The mission is to research global climate changes and their effect on weather phenomena.
Envisat	2002 – present	This is the largest earth observation satellite ever built. The ESA satellite carries 10 sophisticated optical and radar instruments that provide continuous observation and measurement of the earth's oceans, land, ice caps and atmosphere for the study of various natural and man-made contributors to climate change. It also provides the wealth of information needed for the management of natural resources.
ICESat	2003 – present	ICESat is a small satellite mission flying the Geoscience Laser Altimeter System (GLAS) in a near-polar orbit. GLAS measures to the elevation of the earth's ice sheets, clouds and land.

Source: NASA, 2008; ESA, 2008.

already mentioned for soil moisture detection and to be launched in 2009, should allow detection of water salinity to some degree. The US-Argentine Aquarius satellite, to be launched in 2010, should also better measure sea surface salinity (SSS) on a global basis.

Snow cover and sea ice

The cryosphere consists of sea, lake and river ice, snow cover, glaciers, ice caps and ice sheets, and frozen ground (including permafrost). It is an important part of the global climate system, and more than one-sixth of the global population lives in regions that depend on snow and glaciers for water supply. Changes to the cryosphere have significant implications for global sea levels, regional water resources and both terrestrial and aquatic ecosystems. Although 25 years of satellite data are available, the many processes involved in the cryosphere are still poorly understood. One recurring reason mentioned by scientists is the fact that many missions were of short duration and the short revisit time of satellites causes important observations to be missed (EARSEL, 2006). Despite these shortcomings, new data are becoming available and innovative applications are being developed in many countries.

Snow – The amount and timing of snowmelt runoff from snow and glaciers provide important information for the management of water resources, including flood prediction and hydropower operations. Main data sources include satellite data from polar orbiting and geostationary satellites with visible/near-infrared instruments – such as Land (remote sensing) Satellite (Landsat); Moderate Resolution Imaging Spectroradiometer (MODIS); Medium Resolution Imaging Spectrometer (MERIS); Geostationary Operational Environmental Satellites (GOES); and Advanced Very High Resolution Radiometer (AVHRR). For example, MODIS data are being used to monitor the snow cover dynamics of areas greater than 10 square kilometres, together with AVHRR data that provide continuous information on snow cover during daytime in cloud-free areas. Snow of moderate depth can be measured using passive microwave data. These systems have capabilities for day/night monitoring regardless of cloud cover. Optical and radar imagery are today used operationally in the monitoring of snow cover in Scandinavia for hydropower planning and flood prediction.

Sea ice – The safety and efficiency of sea transportation, offshore operations, fisheries and other marine activities identified in previous chapters have provided the motivation to establish sea ice monitoring and snow forecasting services in many countries, in addition to the classic weather services. These services using satellite data (from the SSM/I sensors onboard DMSP satellites, SeaWinds, Envisat, Radarsat) are usually limited to national areas of interest, although international research on the Arctic and Antarctic increasingly calls for the development of new ice monitoring instruments. A number of projects led by international consortia are developing generic sea ice data monitoring to be reused by intermediate users (including commercial value adders) who will then generate services, drawing on other information to serve specific end-users. The European Polar View project for example consists of the development and qualification of customised ice information services, including high resolution ice charts (*e.g.* for Svalbard region) and ice forecasting information services. These are all primarily based on Envisat and Radarsat data.

Table 4.6. **Requirements *vs.* capabilities for sea ice monitoring and snow cover**

Requirements	Capabilities of satellites
Spatial distribution of sea ice	The use of remote sensing for detecting and monitoring icebergs still has its limitations, mainly due to the scale of the targets. Future SAR missions with increased spatial resolution, multiple frequencies and polarisations might solve some of the current problems.
Sea ice thickness and type (first-year and multiyear ice) and their variation with time	Ice thickness is more difficult to monitor than ice extent. With satellite-based techniques only recently introduced, observations have been spatially and temporally limited.
Measurements of snow accumulation and melt	Operational systems are in place in a number of countries, although information about snow depth is sometimes limited.

SeaWinds' Ku-band backscatter data are used to generate daily sea ice maps at 6-kilometre resolution. Such data can be used to monitor seasonal ice changes, and scatterometer data have already been used over the years to track the break-up of giant icebergs.[3] In 1992 a giant iceberg nearly the size of Rhode Island, known as B10A, broke away from the Thwaites glacier in Antarctica. In September 1996 the iceberg was observed floating in the Antarctic ice pack by NASA's Scatterometer (NSCAT) instrument on board Japan's Advanced Earth Observing Satellite (ADEOS). By 1999, conventional methods of tracking icebergs lost sight of B10A's location due to poor visibility and cloudy Antarctic winters. However, it was spotted again by the SeaWinds scatterometer on the Quikscat satellite in July 1999 during its first pass over Antarctica. The iceberg was then breaking up and being driven by ocean currents and winds into the shipping lane, where it posed a threat to commercial, cruise and fishing ships. More recent examples have shown that a combination of optical and radar imagery that is high resolution (10 metres or less) can improve iceberg detection (Vitaly *et al.*, 2008). More satellites are expected to be launched over the next three years with increased sensing capabilities, which should allow better coverage (*e.g.* high resolution radar satellite TerraSar-X, COSMO-SkyMed's four radar satellites constellation).

Water quality

Scientific activities are ongoing to develop new methods to extract a variety of in-water bio-optical properties from remote sensing data. The objective is to be able to monitor the quality of water bodies, detecting in particular natural and man-made pollution. Space-based sensors look primarily at "ocean colour" in waters, which refers to the presence and concentration of specific minerals or substances. The prime observables of space-based ocean colour instruments are the chlorophyll and gelbstoff concentrations in the surface layers. The concentration of chlorophyll is used to estimate the amount of phytoplankton in waters, and hence the abundance of ocean biota, while biologists use ocean colour and chlorophyll-a products to predict harmful algal blooms. The first images of ocean colour distribution were taken from space by Nimbus 7. This satellite, launched in 1978, carried a spectroradiometer instrument called CZCS (Coastal Zone Colour Scanner) (Kramer, 2002). The data of this first generation of ocean colour instruments contributed greatly to understanding the marine environment and its biological, biochemical and physical processes.

Box 4.2. **Monitoring jellyfish proliferation in seas and oceans**

The proliferation of jellyfish (also known as medusae) has been observed in seas and oceans around the world over the past decade, with increasing environmental and economic effects. In November 2007, a dense pack of billions of "mauve stinger" jellyfish, usually found in warm Mediterranean waters, killed about 120 000 salmon overnight in a fishery off the coast of Ireland north of Belfast. In spring 2008, a giant jellyfish native to Australia called Phyllorhiza punctata was detected in the waters of the Gulf of Mexico, up the eastern coast of Florida and as far north as North Carolina (ClimateScienceWatch, 2008). Two centuries' worth of local data show that jellyfish populations tend to swell every 12 years, remain stable four to six years, and then subside. However, a new population growth is expected to reach a peak in the summer of 2008 for the eighth consecutive year. The frequency and persistence of the phenomenon seem to be driven by overfishing (i.e. jellyfish occupy the place of many other species, including their former predators), pollution and climate change (i.e. warmer waters, strong ocean currents, altered ecosystems). Monitoring jellyfish populations, of which 30 000 different species have already been identified, can be performed via acoustic surveys at sea. Another method now in the experimental phase is to use computer hydrodynamic models coupled with sea surface temperature and salinity data from satellites (NOAA, 2001). The objective is to map locations where jellyfish are likely to be found, usually in moderate salinity and warm waters. For example, NOAA provides near-real-time maps of sea nettle distribution in the Chesapeake Bay, the largest US estuary, again on an experimental basis (NOAA, 2008).

Source: NOAA, 2001, 2008; Hay, 2006; ClimateScienceWatch, 2008.

Although there are many water variables that remote sensing can already detect (chlorophyll-a, suspended solids, turbidity), the complexities of water's optical properties still present difficulties in the use of satellite remote sensing for coastal and inland waters. A resolution of 100 metres signifies that the mean water quality can be detected in a pixel of 100 x 100 metres. However, the poor spatial resolution of some sensors limits their use for small lakes and archipelagos, since it becomes harder to distinguish between land and water (Kallio, 2002). Also, two classes that define the quality of optic observations are used in ocean water study. "Case 1 waters" signals that phytoplankton and their derived products are the main influence on the optical field, whereas for "Case 2 waters" there are additional seawater constituents (*i.e.* suspended sediments, dissolved organic matter), which makes it more difficult to analyse the optical field. Coastal and inland waters are usually "Case 2"; here, the current algorithms and space sensors' resolution still need to be improved (Eurico, 2005).

Box 4.3. **Measuring river flow with radar satellite**

The existing method of "streamgaging", developed in the early 1900s, consists of physically measuring the channel geometry and velocity of the water on a periodic basis. Because these data are needed over the entire range of river flow situations, personnel and equipment are often subjected to dangerous weather and flow conditions. In addition, because many of the gaging locations are remote, obtaining the data is expensive and cannot be done frequently or continuously. To measure flow at a given cross-section of a river, two pieces of information are essential: 1) water velocity; and 2) channel cross-sectional area (depth and width). Satellites equipped with radar instruments can be used with some limitations for those two types of measures. Water surface velocity can be measured at various points across the river with radar altimetry (using the Bragg scatter principle of high-frequency pulsed doppler radar signal). Hundreds of separate radar pulses are sent per second from satellites in low earth orbit over bodies of water, and the time it takes their echoes to bounce back recorded. Non-contact methods of streamgaging show great promise, but much remains to be learned. Conductivity has a significant negative effect on radar energy in water, and there are physical limits to the depth of radar energy penetration.

Source: Adapted from Hirsch 2004; Plant *et al.*, 2005.

Remote sensing from space therefore has limitations when it comes to water quality monitoring. Instruments can at present only directly detect some of the pollutants, such as oil, sediment and thermal discharges. The others can in some cases be detected by proxy, but only via ephemeral site-specific correlations between remotely sensed features and the pollutants in question. Thus fertilisers, pesticides, nutrients and colourless toxic and industrial wastes (containing, for example, heavy metals, aromatic hydrocarbons, acid, etc.) cannot be detected, but in some instances their concentrations (measured by field survey at the time of image acquisition) may be correlated with levels of substances that *can* be detected. Such ephemeral relationships may be demonstrated through experiment, but even then they can be a long way from allowing operational, cost-effective detection and monitoring of pollutants (Green *et al.*, 1999).

According to the GCOS group, the best-quality basin-wide or global data on ocean colour are currently being collected using SeaWiFS, the Medium Resolution Imaging Spectrometer (MERIS), and the Moderate Resolution Imaging Spectroradiometer (MODIS), all carried onboard different satellites. There is no guarantee, however, that either of these instruments will still be in operation a decade from now. It is therefore necessary to ensure the continuity of equal or even better-quality ocean colour measurements in the future, to facilitate chlorophyll climatology studies on regional and global scales (CEOS, 2007).

Water temperature

Over the past decade, a large number of satellites' sea surface temperature (SST) products, derived by different groups or agencies from various satellite sensors and platforms, have become available in near-real-time. Data on sea surface temperature maps notably help meteorologists predict weather and fishermen locate prime fishing areas.

The instruments used are all passive sensors (radiometers), which measure the natural radiation emitted by the earth's surface and propagated through the atmosphere. Several space instruments are available today that provide hundreds of data products with a range of spatial, spectral, radiometric and temporal scales (Table 4.7). The Advanced Very High Resolution Radiometer (AVHRR), the Along Track Scanning Radiometer (AATSR), and the Moderate Resolution Imaging Spectroradiometer (MODIS) sensors are the main sources of sea surface temperature data (CEOS, 2006). The AATSR series of instruments have for example delivered a highly accurate time series of sea surface temperature variations over a 17-year period (1991-2008), which has enabled a unique characterisation of global warming during that time (ESA, 2008).

The spatial dimension and spectral resolution of these sensors are not, however, suitable for small-scale measurements. Satellite imagery must be integrated with airborne and ground-based remote sensing systems (*i.e.* hyperspectral sensors with greater resolution) for the documentation of point areas of importance, such as around dams and industries. These systems can be hired on a case-by-case basis, but they cannot produce the temporal resolution of satellite imagery and do not provide comprehensive coverage of large basin systems (ISU, 2004).

Table 4.7. **Overview of certain space-borne sensors for ocean colour detection**

Sensor (sensor provider)	Platform	Ground resolution (km)	Swath width (km)	Spectral bands	Data revisit (days)	Dates of operations
CZCS (NASA, United States)	Nimbus 7	0.825	1 600	5 bands	2	Oct. 1978 – June 1986
OCE (NASA, United States)	Space Shuttle STS-2 (SIR-A mission)	0.90	180	8 bands	NA	Nov. 1981
MOS A, B and C (DLR, Germany)	IRS-P3 (ISRO, India); Priroda (Mir Module, the Russian Federation)	57x1.40; 0.52x0.52; 0.52x0.64	195 200 192	4 bands 13 bands 1 band	NA	March 1996 April 1996
OCTS (NASDA, Japan), POLDER (CNES, France)	ADEOS	0.70	1 400	8 bands	2	Aug. 1996 – June 1997 (some ocean colour products)
SeaWIFS (OSC/Orbimage, United States)	OrbView-2	From. 1.1 to 4.5	From 2800 to 1500	8 bands	2	Aug. 1997 – present
OCI (NSPO, Chinese Taipei)	ROCSAT-1	0.80	691	6 bands	NA	Jan. 1999 – present
OCM (ISRO, India)	IRS-P4 (OceanSat1)	0.36	1 420	8 bands	2	May 1999 – present
MODIS (NASA, United States)	Terra	1	2 330	9 bands	2	Dec. 1999 – present
OSMI (KARI, Korea)	KOMPSAT-1	1	800	8 bands	NA	Dec. 1999 –present
MERIS (ESA, Europe)	Envisat	0.3 and 1.2	1 150	15 bands	3	2001 – present
MODIS (NASA, United States)	Aqua					2001 – present
OCS (SITP, China)	HY-1 (Haiyang-1)					2002 – present
GLI (NASDA, Japan), Polder-2 (CNES, France), SeaWinds	ADEOS-2	1 to 0.25	1 600	36 bands	2	2002 – 2003
COIS (NRL, United States)	NEMO (Navy Earthmap Observer)	0.03 to 0.06	30	210	7 (2.5 in some cases)	2003 – present

Source: Updapted from Kramer, 2002.

As mentioned by Roquet (2006), many efforts are under way at the international and European levels to validate and make available in near-real-time high-quality satellite water temperature products. Much scientific work, including impact studies, remains to be performed to define the best way of assimilating high resolution SST products into different global, regional or coastal ocean models.

Box 4.4. **Coupling essential *in situ* systems and satellites for monitoring sea level rise and ocean temperature**

- *ARGO floats* – In November 2007, the international Argo system reached a significant milestone with the launch of its 3 000th float operating in the ocean. The first Argo float was launched in 2000 to improve estimates and forecasts of sea level rise, climate and hurricanes (UNESCO, 2008). The most obvious benefit from Argo has been a marked reduction in the uncertainty of ocean heat storage calculations (a key factor in determining the rate of global climate warming and sea level rise). The steady stream of Argo data, together with global-scale satellite measurements from radar altimeters, has also made possible advances in coupled ocean atmosphere models. These have led to seasonal climate forecasts and routine analysis and forecasting of the state of the subsurface ocean. Argo data are also being used in an ever-widening range of research applications that have provided new insights into how the ocean and atmosphere interact in extreme as well as normal conditions. Maintaining the array's size and global coverage in the coming decades is Argo's next challenge. The Argo implementation programme was designed assuming for a lifetime of four years with an anticipated failure rate of 10% over that period. A constant 3 000-float array, therefore, requires 825 floats to be deployed annually.

- *Expendable bathythermographs (XBT)* – Expendable bathythermographs are devices sent from vessels to obtain information on the temperature structure of the ocean to depths of up to 2 000 metres. A small probe dropped from the ship measures the temperature as it falls through the water. By plotting temperature as a function of depth, scientists can put together a temperature profile of the water. Many different types of vessels of opportunity can be employed (*i.e.* ferries, cargo ships). For example, the NOAA has a dedicated XBT programme (SEAS) with about 80 voluntary ships. Observations from these vessels are collected and coded using the World Meteorological Organisation's bathy report format, and transmitted via the GOES and INMARSAT C satellites. Those 80 SEAS vessels produce more than 14 000 XBT observations each year.

Source: GODAE, 2007; NOAA, 2008; UNESCO, 2008.

Altimetry, geopotential height and topography

A number of satellite instruments are unique and comprehensive sources of altimetry, geopotential height and topographic data.

Altimetry from space – Altimetry is a technique used to measure height or elevation. For more than 17 years, satellite altimetry has been used to measure sea surface heights. The Franco-American mission Topex/Poseidon and ERS-1 not only demonstrated that space altimetry could work with high precision

(3 centimetres at basin level), but also provided unexpected information for monitoring oceanic phenomena (*e.g.* variations in ocean circulations such as the El Niño 1997-98 event, seasonal changes in oceans, tide mechanisms).[4] There are as of June 2008 five satellites available with altimetry instruments (Jason-1, Jason-2, Envisat, ERS-2, and the Ice, Cloud, and Land Elevation Satellite or ICESat). Continuing operational observations, via follow-up systems, should allow observation of decennial oscillations of the Atlantic and Pacific Oceans and the ongoing global rise of oceans. Those satellites are complemented by observations made using instruments placed on different multipurpose space-based systems (*e.g.* ERS, Radarsat). Initial work over a handful of large water targets has expanded to the current capability to monitor, if sometimes imperfectly, thousands of river and lake heights worldwide. As an example, ICESat, launched in 2003, carries the Geoscience Laser Altimeter System (GLAS). Its purpose is to measure the earth's polar ice sheet mass balance, cloud and aerosol heights, as well as land topography and vegetation characteristics, in some cases with sub-metre height resolution. Continuation of precision altimetry data recording is a prime concern expressed by scientists, mentioned during the 2007 International Ocean Surface Topography Science Team Meeting (Fu, 2007). That recording is key to monitoring and understanding global ocean circulation and sea level variability in relation to global climate variability. The NOAA-EUMETSAT Jason-3 satellite, NASA's Surface Water and Ocean Topography (SWOT) satellite, the AltiKa satellite, and ESA's Sentinel 3 are the next major missions planned for extending the climate data record.

Geopotential height – Aside from knowing the geometric height of a body (the elevation above mean sea level) – via altimetry, for example – another height measure used in meteorology and climate studies is geopotential height. This is the height of a pressure surface above mean sea level.[5] Since the 1940s, radiosondes have been used to measure pressure, temperature and humidity profiles. A geopotential height is then calculated by combining those data into hydrostatic equations (Haimberger, 2007). As mentioned by Jeannet, Bower and Calpini (2008), there have been tremendous advances in geopotential height measurement over the two last decades. This is partially due to newer radiosondes that use GPS signals to measure geometric height directly and convert it with high precision to geopotential height. This relatively recent use of GPS represents a real technology leap, it brings greatly improved accuracy and standardisation of measures, even if adjustments are often necessary. For example, at Mauritius, all GPS height measurements agreed on average to within ±20 metres from the surface to a 34-kilometre altitude, whereas mid-80s technology provided height measurement differences in the order of 500 metres to a 30-kilometre altitude (Jeannet, Bower and Calpini, 2008).

Topography and geomorphology (shape of the earth) – Satellite sensors can provide effective topographic and geomorphologic data. This is particularly

Box 4.5. **Tracking the world's water supplies**

Launched in March 2002, the Gravity Recovery and Climate Experiment (GRACE) is a five-year mission, still ongoing as of June 2008, to better understand the earth's gravity field. Using a pair of roving satellites 220 kilometres apart – GRACE-1 and –2 – water supply changes are measured around the world. Even if the water is captured in snow, rivers or underground aquifers, the satellites can detect the mass and trace its progress. GRACE has only been tracking world water supplies for three years, so the data cannot yet be used to determine where a water problem may emerge next, however much has been learned.

Evidence indicates that ground water is being depleted in the central valley of California, parts of India and in the Nubian Valley in Africa. The annual 21.6 millimetre water shrinkage deep in the Congo translates to 260 cubic kilometres of water lost between 2003 and 2006. Meanwhile, GRACE data show that the Nile has been receding an average of 9.3 millimetres a year while the Zambezi has receded 16.3 millimetres (the measurements take into account water flowing in the river, that in the soil and ground water). Africans consume about 136 cubic kilometres of drinking water a year, so the losses in the Congo over a three-year period are, very roughly, equivalent to two years' worth of drinking water. Natural climate variation, however, can raise or lower water in a given period. During the same period, the Colorado River in the United States rose by an average of 37.5 millimetres a year.

Source: Kanellos, 2006.

useful for detecting water movements (surface and ground water) and where water is stored (aquifers, surface waters). One of the methods most widely used to create three-dimensional digital elevation models (DEMs) is radar measurement (*e.g.* Radarsat). At present, most of the information is obtained primarily from multi-band optical imagers and synthetic aperture radar (SAR) instruments with stereo image capabilities (*e.g.* Radarsat, Envisat). The pointing capability of some optical instruments allows the production of stereo images from data gathered on a single orbit (*e.g.* by ASTER) or multiple orbits (*e.g.* by SPOT series); these are then used to create digital elevation maps that render a more accurate description of terrain. The ability of SAR systems to penetrate through the cloud cover and the dense plant canopies makes the technique particularly valuable in rainforest and high-latitude boreal forest studies. Instruments such as Advanced SAR (ASAR) and Phased Array type L-band Synthetic Aperture Radar (PALSAR) will provide data for applications in agriculture, forestry, land cover classification, hydrology, and cartography.

This brief overview of satellite requirements and capabilities for climate monitoring has shown that space technologies – whether earth observation or

90

navigation systems – are means to obtain unique types of data. Still, there are some clear technical and governance-related limitations (*e.g.* obtaining real-time data, the length of R&D missions and their transformation into long-term operational missions) that will be explored later in this report.

Controlling maritime and marine areas

As discussed in the previous chapters, a number of countries are increasingly controlling their "seas", whether for fish stock assessment and protection, offshore development, pollution management or immigration control. One constant is the growing requirement to monitor and control large maritime and marine areas. A wide variety of tools are already available to countries for doing so, often using space-based signals or data. There are two types: co-operative surveillance tools (reporting on voice via VHF radio, Automatic Identification Systems, Long-Range Identification and Tracking, fisheries vessel monitoring systems) and non-cooperative surveillance tools (coastal radar, radar satellites).

Co-operative surveillance tools

"Co-operative" systems imply the setting up of electronic devices (transceivers) on board vessels, often at the shipmaster's own cost. Ships of certain sizes or following certain sea routes are already obliged under international law to carry special communications equipment.

Automatic identification system (AIS)

The automatic identification system was developed primarily to improve maritime safety by assisting the navigation of ships.[6] AIS is a ship-borne transponder broadcasting ship, voyage, and safety-related reports via VHF; the only satellite contribution is GPS positioning (Cauzac *et al.,* 2008).

A coastal country can use information from passing vessels to obtain an almost real-time maritime picture of its home waters. It needs to establish a VHF receiving chain along the coast for AIS signals, the range of which can typically be 40–50 kilometres out from the coast. AIS serves in a vessel-to-vessel mode for collision avoidance. It is both a means for coastal states to obtain information about a ship and its cargo, and a ship-to-shore method for vessel traffic management. The advantage of AIS is that ships can be alerted to the presence of other ships – and coastal authorities can pinpoint the position of ships – that may be undetectable by traditional radar (*e.g.* because of the harsh topography of certain shorelines).

AIS has been mandatory on all new ships in international traffic since 1 July 2002. It has covered all passenger ships, tankers and other ships of 300 tons engaged in international voyages since the end of 2004. It should be fully implemented in late 2008 and include all ships of 500 tons or more in national voyages. But national and regional sharing of AIS data is already developing fast: for example, Europe-wide sharing of vessel traffic data is progressing under SafeSeaNet, based on a 2002 European Community directive (European Commission/Joint Research Centre, 2008).

Long-Range Identification and Tracking (LRIT)

In addition to AIS, all passenger ships and all cargo ships over 300 gross tons need also to carry on board by the end of 2008 (for the most recent ships) a Long-Range Identification and Tracking device, using satellite signals to carry back information to ground receivers (IMO, 2006b). This LRIT system allows coastal countries to receive information about ships navigating within 1 000 nautical miles off their coast (their identity, location and the date and time of the position). According to international law, LRIT data will be available only to authorities entitled to receive such information, since confidentiality issues are included in the regulatory provisions. Each state or regional authority will be able to access an international LRIT database, but no global interface is planned with the LRIT and AIS broadcast identification. This is partially to protect commercial confidentiality but also because the technical challenges are huge, as will be shown in the next section.

One of the more important distinctions between LRIT and AIS, apart from the obvious one of range, is that AIS is a broadcast system that sends information easily intercepted by anyone, not only coastal authorities. Data derived through LRIT will, as stated above, be available only to proper recipients. SOLAS contracting governments will be entitled to receive information about ships navigating within a distance not exceeding 1 000 nautical miles off their coast (IMO, 2006a).

Vessel monitoring system (VMS)

Fishing is by its very nature an activity that relies on spatial information inputs. The introduction of GPS devices in commercial fishing and earth observation imagery, coupled with plotters, was a major advance in fishing management. But the development of dedicated fishing vessel monitoring systems (VMS) was a real step forward in improving the monitoring and control of fisheries' activities (Figure 4.3).

Fishing vessel monitoring systems are relatively complex; they rely on equipment installed on fishing boats to provide information about the vessels' position and activity, often using satellite communications and navigation (FAO, 1999). These devices automatically send data to a satellite system that transmits them to a land base station; the station in turn sends them to the appropriate Fisheries Monitoring Centre (FMC), where the information is cross-checked with other data. The effectiveness of the system depends on the

Figure 4.3. **Vessel monitoring system (VMS) for EU's Common Fisheries Policy**

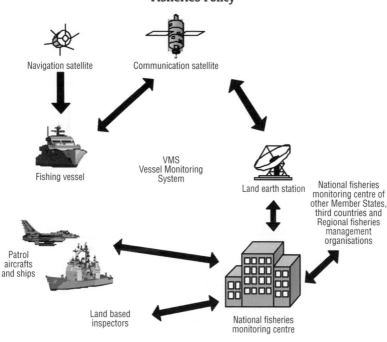

Source: EC, 2007.

reactivity of the different controlling components, which include in particular land-based inspectors and patrol vessels and aircrafts. Monitoring systems have been already in use for more than twenty years in some countries (Cauzac et al., 2008) and are developing around the world. In Europe for example, national and European Community funding was made available in the early 1990s for pilot-projects, which demonstrated the reliability of satellite position monitoring and its potential for promoting the effectiveness of existing surveillance resources (European Commission / Joint Research Centre, 2008).

Non-cooperative surveillance tools

Non-cooperative surveillance systems have been increasingly used, with some success, by national authorities to check activity in their maritime and marine zones. Many countries have installed coastal ocean radars, particularly in connection with developing their Vessel Traffic Services (Box 4.6). The range available for ship detection varies greatly depending on the radar systems, from a few kilometres to more than 300 kilometres depending on weather conditions.

Box 4.6. **Description of Vessel Traffic Services (VTS)**

The purpose of a Vessel Traffic Service is to provide active monitoring and navigational advice for vessels in particularly confined and busy waterways. There are two main types of VTS: surveilled and non-surveilled. Surveilled systems consist of one or more land-based sensors (i.e. radar, AIS and closed circuit television sites), which output their signals to a central location where operators monitor and manage vessel traffic movement. Non-surveilled systems consist of one or more reporting points at which ships are required to report their identity, course, speed and other data to the monitoring authority. They both encompass a wide range of techniques and capabilities aimed at preventing vessel collisions, rammings, and groundings in the harbour, and assisting harbour approach and the inland waterway phase of navigation. They are also designed to expedite ship movements, increase transportation system efficiency, and improve all-weather operating capability.

The VHF-FM communications network forms the basis of most major services. Transiting vessels make position reports to a vessel traffic centre by radiotelephone and are in turn provided with accurate, complete and timely navigational safety information. The addition of a network of radars and closed circuit television cameras for surveillance and computer-assisted tracking, similar to that used in air traffic control, allows the VTS to play a more significant role in marine traffic management. This decreases vessel congestion, the number of critical encounter situations, and the probability of a marine casualty resulting in environmental damage.

Source: Adapted from US Coast Guards, 2008 and IMO, 2006a.

A relatively new tool for surveillance consists of complementary optical and radar satellite data; it is used in particular to identify ships and oil spills (Greidanus, 2008). Carried onboard the Radarsat and Envisat satellites for example, radar sensors allow day and night identification. Small vessels down to 15-30 metres can be detected in the images; this however depends on weather conditions, since the detection capabilities of ship targets are greatly influenced by the wind speed and wind direction in relation to the track of the satellite (Konsberg Satellite Services, 2008). Furthermore, although radar satellite surveillance provides wide area coverage, it is often limited by revisit times and by the time necessary to process the imagery, analyse it and exploit it (Tunaley, 2004). Therefore, it is a combination of both optical and radar imagery that allows better detection and ship recognition (Table 4.8).

Radar satellite data are also used for oil spill detection, especially because of their extensive coverage and all-weather/all-day capabilities (Wahl T., 1996; Serra-Sogas et al., 2008). Distinguishing real oil spills from lookalike features (known as "false positives") is the biggest challenge of this technique (Brekke

Table 4.8. **Radar and optical satellite images for ship detection**

	Radar satellite images	Optical satellite images
Level of detail	Generally low resolution (8 to 50 metres)[1]	High resolution (< 1-10 metres)
Geographic coverage	Wide area (up to 400 kilometres areas)	Small area (10-60 kilometres)
Particular conditions	Independent of clouds, night (24h vision)	Daytime, clear skies
Main use	Use for detection	Use for recognition

1. Although new radar satellites are progressively becoming fully operational and can provide more detailed resolution (*e.g.* CosmoSkymed, TerraSARX).

and Solberg, 2005). According to a number of validation experiments using onsite airborne observations, the overall probability of successful oil detection by satellite varies from 40% to 60% (MarCoast, 2006). Different detection methods are constantly being refined, improved and benchmarked by the scientific community. But this is still very much work in progress. One lesson learned from the many demonstration projects under way around the world is that acquiring oil slick data can take a long time. A good illustration is Synthetic Aperture Radar (SAR) data from the European satellite ENVISAT. The MarCoast service guarantees a delivery time of 30 minutes after the satellite overflight, and while the SAR-IM instrument allows SAR image resolution of 30 metres in all weather and day/night detection capabilities, the frequency of passage over Mediterranean waters ranges from two to three days (Ferraro *et al.*, 2006). Another aspect is that delays may be exacerbated by conflicts between the needs of different user communities. For instance, the wide swath modes that are optimal for oil slick detection work may not offer a resolution high enough to detect small vessels. The benefits of such monitoring systems may be increased by the use of new SAR systems (*e.g.* Cosmo-Skymed, SAR-Lupe, Radarsat-2) and by possible future deployment of constellations of smaller, single-purpose satellites alongside the existing large multipurpose ones. Finally, more work is needed to calibrate the different sources of data useful to users, since there can be discrepancies between satellite ship detection reports via imagery and the AIS records in the order of several hundreds of metres (Cauzac *et al.*, 2008).

To improve the overall detection system, there is the need for co-ordination between satellite overpasses and aerial surveillance flights (Brekke and Solberg, 2005). In addition, from the users' point of view, one of the most useful elements for future systems is the integration of AIS vessel information and oil slick drift forecasting data, so as to identify vessels located in the polluted area and trace back the polluter (Figueroa *et al.*, 2008). More information on the detection of man-made pollution from space and the derived benefits is presented in Chapter 6.

Table 4.9. **Requirements vs. capabilities for monitoring oil pollutions**

Requirements	Capabilities of satellites
Geographic: A rapid survey of pollutants over large or remote areas which could not feasibly be surveyed from the surface	Only satellites can provide global coverage. A few radar satellites currently provide data useful for detecting large oil spills.
Timescale: Ideally daily surveillance for a strong deterrence + rapid data availability (in less than 1 hour)	As an example, the frequency of data acquisition on average (taking into account all available satellites) can be assumed to be about 2-3 days over the Mediterranean. Relative to equatorial waters, SAR coverage is 1.2 times as frequent for the Mediterranean Sea, 2.0 times more likely for the North Sea, 3 times for the Norwegian Sea and 4 times for the Barents Sea (MarCoast, 2006). Increased coverage will necessitate more satellite SAR sensors.
Oil spill scope: A synoptic overview of the pollution and clear demarcation of its boundaries	Since the pollutants can only be detected and measured if they can be differentiated from other substances in the water, sensors of high spectral resolution can provide information.
Monitoring: Once reliable calibration has been achieved, the potential for measurement of the pollutant without the need for more field survey	Again, the revisit times are currently too long to allow real-time monitoring.
Access to data: Results available through web systems operated at regional or at country level	Ongoing efforts in a number of countries to provide operational data (*e.g.* Canada, United States, France, Italy, Norway).

Table 4.10. **Civilian/dual-use radar satellites in operation**
(June 2008)

	Planned design life	Radar imaging frequency	Spatial resolution	Status
Radarsat-2	7 years	C-Band	3 to 100 metres	In operation since Dec. 2007
Radarsat-1	5 years	C-Band	10 to 100 metres	In operation since 1995
Envisat ASAR	5 years	C-Band	30 to 1 000 metres	In operation since 2002
ALOS PALSAR	5 years	L-Band	10 to 100 metres	In operation since 2006
Cosmos Skymed (4 satellites)	5 years / satellite	X-Band	1 to 100 metres	3 satellites launched (2 in 2007, 1 in 2008), next in 2009
TerraSAR-X	5 years	X-Band	1 to 15 metres	In operation since June 2007

Source: Adapted from Serra-Sogas *et al.*, 2008.

Safety at sea

The ability to communicate instantaneously from anywhere in the world and navigate with increased precision has revolutionised maritime transport activities. The use of satellites and the maturing of information technology equipment have allowed a number of relatively new commercial systems to be established in recent years, increasing safety at sea but also the requirements for ship owners and operators.

International requirements to increase sea safety

With the growth in global sea trade, governments have increasingly co-operated via the International Maritime Organisation (IMO) to improve safety at sea, developing in particular regulations pertaining to the use of new technological capabilities. The International Convention for the Safety of Life at Sea (SOLAS) Convention is today regarded as one of the most important international treaties dealing with the safety of ships, particularly merchant ships (IMO, 2006a). The first version was adopted in 1914 in response to the Titanic disaster; the latest Convention dates back to 1974 and is amended regularly, particularly in relation to new technological capabilities (IOC/UNESCO, 2005).[7] Since the late 1990s, diverse equipment – most of it using satellite telecommunications and navigation – is required on board ships.

Global Maritime Distress Safety System (GMDSS)

A hundred years after the first use of wireless technology to aid a ship in distress using the Morse code, the Global Maritime Distress Safety System became operational in 1999. It is an international integrated communications system using satellite and terrestrial radio communication to ensure that no matter where a ship is in distress, aid can be dispatched.[8] Under the SOLAS Convention, all passenger ships and all cargo ships over 300 gross tons on international voyages, including commercial fishing vessels, have to carry GMDSS-approved equipment for sending and receiving distress alerts and maritime safety information, and for general communications. Among these devices are satellite emergency positioning indicating radio beacons (EPIRB), designed to operate with the international satellite-based search and rescue system Cospas-Sarsat (Box 4.7).

This GMDSS adoption has spurred the development of maritime markets for space applications, as Chapter 5 will demonstrate. However, because of the many vessels around the world that are not subject to the Convention, in particular small commercial and leisure crafts, there is an important international requirement to maintain some of the radio procedures used prior to the introduction of the GMDSS. That way a common means of communication is available between all classes of vessels for distress and safety purposes. Building on the Global Maritime Distress Safety System, new international norms have been recently adopted, spurring again the development of a new generation of commercial communications and navigation equipment.

Ship Security Alert System (SSAS)

As piracy and terrorist attacks have increased over the past decade, the International Maritime Organization (IMO) has defined a set of mandatory requirements to improve security for ships: the International Ship and Port

---

Box 4.7. **The Cospas-Sarsat system**

The Cospas-Sarsat system was set up in 1982. The acronyms stand for "Space system for the search of vessels in distress" (Cospas in Russian) and "Search and rescue satellite-aided satellite tracking system" (Sarsat). The programme is the result of international collaboration by the United States, France, the Russian Federation and Canada. It provides an alert and satellite positioning aid function for the search and rescue of persons in physical distress, on land or at sea anywhere in the world. The system uses instruments onboard satellites in geostationary and low-altitude earth orbits that detect the signals transmitted by distress radio beacons and ground receiving stations that receive and process the satellite downlink signal to generate distress alerts to rescue centres throughout the world. More than 900 000 maritime, aviation, and land-based distress beacons use the system worldwide. Cospas-Sarsat is credited with assisting search and rescue forces in the saving of more than 18 500 lives since the first satellite launch in 1982.

Source: Cospar Sarsat, 2007.

Facility Security Code (ISPS). This code, which was signed in 2002 and came into force in late 2004, aims to establish an international framework for detecting and assessing security threats and taking preventive measures against security incidents affecting ships or port facilities used in international trade.[9] It has already contributed to enhancing international maritime security (OECD, 2006). The regulation stipulates that a large range of ships have to be equipped with Ship Security Alert System (SSAS). The system consists of two alarm buttons that can be activated in case of piracy or a terrorist attack, and send an alert via telecommunications satellites (usually the Inmarsat, Argos or Iridium systems). The ships that need SSAS on board include all passenger ships and all cargo vessels of 500 gross tons and upwards, particularly oil and chemical tankers, gas carriers, bulk carriers, and high-speed cargo.

Technologies and equipment

As seen in the previous section, today certain communications and navigation equipment are mandatory onboard ships. Often these devices use satellites as the backbone infrastructure for navigation signals and data communications. In parallel with mandatory apparatus, other equipment has been developed for more cost-efficient sea navigation and has more generally facilitated life onboard (*e.g.* satellite phone, Internet).

Communications services at sea and in remote areas depend mainly on commercial satellites that are in geostationary orbits (Inmarsat systems) and/or constellations in low earth orbits (Iridium, GlobalStar, OrbComm).[10]

Box 4.8. **Improving alerts in case of tsunamis via satellite communications links**

In December 2007, the Intergovernmental Oceanographic Commission of UNESCO (IOC) signed an agreement with Inmarsat, provider of global mobile satellite communications, to further upgrade and enhance the Indian Ocean Tsunami Warning System. Under the agreement, Inmarsat will provide Broadband Global Area Network (BGAN) transmission service to 50 sea level stations in the Indian Ocean. BGAN will enable transmission of sea level observation every minute, *versus* the current system that uses meteorological satellites to transmit data every 15 minutes. Time saved by faster transmission represents significant progress; in the eastern and northeastern Indian Ocean, a tsunami wave can hit the shore in about 30 minutes. Increasing the transmission frequency will provide more time and information for national warning authorities to alert coastal populations at risk.

Source: IOC/UNESCO, 2007b.

For navigation and localisation applications, signals from the GPS satellites' constellation are used the most, although other navigation systems (Glonass and Egnos, precursor to the European Galileo) and local and regional GPS augmentation systems (AWAAS) exist. The Argos satellite-based location and data collection system, using signals sent from dedicated instruments carried onboard a number of meteorological satellites, is also particularly relevant for a large number of applications and scientific research.[11]

Most commercial equipment that has been licensed for GMDSS, SSAS or AIS includes additional fleet tracking software, to enhance Vessel Monitoring Systems (VMS) and provide fleet tracking possibilities. As the equipment market has rapidly developed, companies tend to provide to the extent possible competitive integrated solutions, not only with the basic on board mandatory communications equipment, but also with extra data sensors to send back to a fleet's owner (ship speed, its heading, but also meteorological indicators at its location: air pressure, waves, wind). For example, the ShipLoc company provides a product that includes a GPS receiver (to calculate the position, heading and speed of the ship) and an Argos transmitter (to relay intermittently the information via satellite), but also the Argos satellite airtime for transmission and SSAS alert service, with continuous monitoring by operators (ShipLoc, 2008).

The standard suite of equipment that ships are required to carry (Table 4.11) depends mainly on the size of the ship and the routes taken. As the requirements for communications and navigation capabilities have grown rapidly in one decade (GMDSS, SSAS, AIS, LRIT), more equipment is coming onto the market.

Table 4.11. **Communications and navigation equipment carried onboard ships**

Equipment	Description
HF, MF and VHF radio installations	Terrestrial (non-satellite) VHF, MF and HF marine radio systems. IMO and ITU now both require that the Digital Selective Calling (DSC)-equipped VHF, MF and HF radios be externally connected to a satellite navigation receiver. That connection will ensure that accurate location information is sent to a rescue co-ordination centre if a distress alert is ever transmitted.
SART (Search and rescue transponder)	Self contained, portable and buoyant radar transponder (receiver and transmitter). SARTs operate in the 9 GHz marine radar band, and when interrogated by a searching ship or aircraft's radar, respond with a signal which is displayed as a series of dots on a radar screen. Although SARTs are generally designed to be used in lifeboats, they are often deployed on ships (two required for ships 500 gross tons or more; one required for ships of between 300 and 500 gross tons).
NAVTEX	Narrow band direct printing telegraphy (fax-like). It is an international, automated system for instantly distributing maritime navigational warnings, weather forecasts and warnings, search and rescue notices and other urgent information to ships.
Satellite emergency positioning indicating radio beacon (EPIRB)	International automatically activated radio emitter, designed to operate with the international satellite-based search and rescue system Cospas-Sarsat. It transmits to a rescue co-ordination centre identification and accurate location of the vessel in distress (increasingly using GPS receivers) from anywhere in the world.[1]
Global positioning system (GPS) device	Satellite-aided measurement system that enables a ship to locate its position anytime, anywhere, expressed in degrees and minutes of latitude and longitude.

1. System developed by Canada, France, the Russian Federation and the United States.
Source: Adapted from US Coast Guards, 2008.

Generally, the closer one navigates to the shore – and is in contact with shore-based very high frequency (VHF) and medium-frequency (MF) radio stations (often absent on the coasts of developing countries) – the less equipment is required. Since most commercial ships tend however to operate in ocean areas, within Inmarsat satellites coverage (below 70 degrees North Latitude and above 70 degrees South Latitude), they need to carry adequate satellite-based equipment. For those navigating outside Inmarsat coverage area, particularly near the poles (above 70 degrees N Latitude and below 70 degrees S Latitude), these ships must in addition be equipped with a high frequency Digital Selective Calling "HF DSC" installation. Existing communications and navigation equipment tends to be more and more integrated. For example, the International Convention for the Safety of Life at Sea now requires that Inmarsat C radio equipment have an integral satellite navigation receiver, or be externally connected to a GPS receiver for precise marine tracking purposes. In addition, the international rules for collision avoidance at sea (COLREGs) are being reviewed regularly to minimise human error, and recent maritime equipment includes GPS receivers to support piloting crews in minimising navigational errors (Statheros, Howells and McDonald-Maier, 2008).

For the sailors, satellite navigation applications offer a number of advantages, in particular with regard to voyage planning facilities, with indications such as range and bearing, estimated time of arrival (ETA) or heading to steer. Current systems can interface with a variety of equipment including printer, plotter, ground-based Loran-C navigation systems, autopilot, satellite communications and navigation computer.[12] The "autolocate" functions usually allow the pilot to compute a satellite fix anywhere in the world when the approximate position is not known within 60 nautical miles.

The backbone for communications at sea, telecommunications satellites are very effective for broadcasting over large areas – as demonstrated by the commercial success of television programming via satellite – but until recently they were not as efficient for two-way communications. While satellites are able to compete very effectively with terrestrial broadcasting services, they tend to be more costly to use for two-way communications over land areas. Moreover, latency has often reduced their ability to be used for services requiring instant interactivity.

A number of commercial satellite communications constellation companies are operational today (Table 4.12). Inmarsat was for decades the main provider of communications at sea, first as an international organisation and then

Table 4.12. **Commercial satellite communications constellations**

Company	Description and Status (May 2008)
Iridium	Iridium is a mobile satellite services provider, which operates a constellation of 66 satellites in low earth orbit. These satellites are expected to continue providing full voice and data services until 2013, by which time a second generation constellation should be in place (signature of a development contract to renew the fleet expected in mid-2009).
GlobalStar	Globalstar operates a constellation of 40 satellites in low earth orbit, offering voice and data services to users in more than 120 countries. Globalstar's products include mobile and fixed satellite telephones, simplex and duplex satellite data modems and service packages, although many Globalstar satellites are experiencing an anomaly resulting in degraded performance for two-way voice and data services at certain times. Globalstar's second-generation satellite constellation, scheduled to be launched beginning in the second half of 2009, should offer advanced wireless voice and high-speed IP Multimedia Subsystem (IMS) services.
OrbComm	ORBCOMM is a satellite data communications company focused exclusively on machine-to-machine (M2M) communications. ORBCOMM provides two-way data communications services around the world through a global network of 29 satellites in low earth orbit and accompanying ground infrastructure. ORBCOMM's transmitters are installed on trucks, trailers, railcars, containers, heavy equipment, fluid tanks, utility meters, pipelines, marine vessels, oil wells and other assets. The system can send and receive short messages, between six bytes and several kilobytes, in near-real-time, allowing users to access critical information readily, often from areas beyond the geographic reach of terrestrial systems.

Source: Iridium, OrbComm and GlobalStar corporate websites, 2008.

as commercial provider. Since its days as a specialised intergovernmental organisation, Inmarsat has provided universal services under an agreement with the IMO. The main technical challenge for the future is the development of new generations of communications satellite offering onboard processing and spot beams, as well as technical standards that will facilitate scalability and bring down costs, notably those of terminals.

Notes

1. The 2007 report from the GCOS *Systematic Observation Requirements for Satellite-based Products for Climate* summarises the main inputs of space instruments and contributes to ongoing work conducted at the UNFCCC and the Intergovernmental Panel on Climate Change.

2. CRYOSAT was ESA's first Earth Explorer Opportunity satellite. The 2004 mission was to determine variations in the thickness of the earth's continental ice sheets and marine ice cover. Its primary objective was to test and quantify the prediction of thinning polar ice due to global warming. After a launch failure in 2004, it was decided to re-launch a similar mission.

3. Data digitalisation has had far-reaching effects on the way satellite data can be used. As an example, NOAA developed in the mid-1990s an Interactive Multisensor Snow and Ice Mapping System to provide snow and ice charts, using geographic information systems (GIS) technology. Before the system was put into operation in 1997, snow and ice charts were constructed manually once a week. Today charts are produced daily, integrating near-real time data when available (Helfrich *et al.*, 2007).

4. In March 2006, an international symposium was organised to take stock of "Fifteen Years of Progress in Radar Altimetry".

5. Geopotential height data are often represented on weather maps by isobar lines connecting points of equal or constant pressure height.

6. Automatic identification systems became mandatory on ships via Resolution 6 in the International Convention for the Safety of Life at Sea (SOLAS).

7. The world's oceans have been divided into thirteen search and rescue areas by the IMO, and in each area the countries concerned have delimited search and rescue regions for which they are responsible.

8. The GMDSS was developed by the International Maritime Organization (IMO) – the specialised agency of the United Nations with responsibility for ship safety and the prevention of marine pollution – in close co-operation with the International Telecommunication Union (ITU) and other international organisations. The latter include notably the World Meteorological Organization (WMO), the International Hydrographic Organization (IHO) and the Cospas-Sarsat partners.

9. Common crimes at sea may include: illegal fishing and reloading, illegal transport of goods, illegal immigration, and illegal dumping.

10. The availability of commercial satellite communications services in any particular country may be subject to government approval.

11. The Argos system enables scientists to gather information on any "object" equipped with a certified transmitter, anywhere in the world – in the oceans, deserts or polar

regions. Argos transmitters come in a variety of shapes and sizes, depending on their purpose. Their messages are recorded by a constellation of satellites carrying Argos instruments, and then relayed to dedicated processing centres. This system has been operational since 1978, and was initiated jointly by France and the United States. Its participants include those countries, India and Europe via Eumetsat.

12. The LORAN (LOng RAnge Navigation) system is a terrestrial radio navigation system involving low frequency radio transmitters that use multiple ground stations to determine the location and/or speed of the receiver. First available during the Second World War and upgraded several times since then, Loran is still used in many countries, as a complementary system to other forms of electronic navigation, including satellites.

Bibliography

Alverson, Keith (2007), "Why the World Needs a Global Ocean Observing System", *Marine Scientist*, No. 21, pp. 25-28.

Brekke, Camilla and Anna Solberg (2005), "Oil Spill Detection by Satellite Remote Sensing", *Remote Sensing of Environment*, Volume 95, No. 1, pp. 1-13.

Brown, Christopher *et al.* (2005), "An Introduction to Satellite Sensors, Observations and Techniques" in R. Miller (ed.), *Remote Sensing of Coastal Aquatic Environments: Technologies, Techniques and Applications*, Springer, Dordrecht, Netherlands.

Cauzac, J.P. and J.Y. LeBras (2008), "What Space Technologies Have Really Changed In Maritime Security", *Proceedings Space Applications 2008*, Toulouse Space Show, 23-25 April.

ClimateScienceWatch (2008), *Non-native Jellyfish Wipe Out Salmon Fishery in Northern Ireland – Another Warning Sign?*, ClimateScienceWatch Website: *www.climatesciencewatch.org/index.php/csw/details/jellyfish_salmon_wipeout*, accessed May.

Cospas-Sarsat (2007), *Cospas-Sarsat System Data*, Report No. 33, December.

CEOS (Committee on Earth Observation Satellites) (2006), *Satellite Observation of the Climate System: The Committee on Earth Observation Satellites (CEOS) Response to the Global Climate Observing System (GCOS) Implementation Plan (IP) 2006*, September.

EARSEL (European Association of Remote Sensing Laboratories) (2006), *Newsletter*, No. 66, June.

ESA (European Space Agency) (2006), ESA Living Planet Programme Strategy Workshop, Frascati, 15-16 February.

ESA (2008), "Europe Celebrates Its First Maritime Day", ESA Press Release, Paris, 20 May.

Eurico, J. D'Sa and Richard L. Miller (2005), "Bio-Optical Properties of Coastal Waters", in R. Miller (ed.), *Remote Sensing of Coastal Aquatic Environments: Technologies, Techniques and Applications*, Springer, Dordrecht, Netherlands.

European Commission/Joint Research Centre (2008), "Integrated Maritime Policy for the EU", Working Document III on maritime surveillance systems, European Commission / Joint Research Centre, Ispra, Italy, 8 January.

FAO (Food and Agricultural Organization) (1999), *FAO Technical Guidelines for Responsible Fisheries: Fishing Operations 1, Supplement 1-1: Vessel Monitoring Systems*, FAO, Rome.

Ferraro, Guido *et al.* (2006), "Satellite Monitoring of Accidental and Deliberate Marine Oil Pollution", Chapter 4 in *Marine Surface Films*, Springer Berlin Heidelberg.

Figueroa, Araceli Pi *et al.* (2008), "Oil Spill and Water Quality Monitoring Spanish Services in MarCoast", Proceedings of Space Applications 2008, Toulouse Space Show, 22-25 April.

Fu, L.-L. (2007), "Report of the 2007 OSTST Meeting", International 2007 Ocean Surface Topography Science Team (OSTST) Meeting, Wrest Point Hobart, Australia, 12-15 March.

Gaston, Robert and Ernesto Rodriguez (2008), *QuikSCAT Follow-On Concept Study*, NASA Jet Propulsion Laboratory, Pasadena, California, JPL Publication 08-18, April.

GCOS (Global Climate Observing System) (2005), *Implementation Plan for the Global Observing System for Climate in Support of the UNFCCC*, GCOS-92, Geneva.

GCOS (2006), *Systematic Observation Requirements for Satellite-Based Products for Climate*, GCOS-107, WMO/TD No. 1338, Geneva.

GODAE (Global Ocean Data Assimilation Experiment) (2007), *IGST XII Meeting Report*, St. John's, Newfoundland, Canada, 7-9 August.

Green, Edmund P. *et al.* (1999), *The Remote Sensing Handbook for Tropical Coastal Management*, UNESCO and the United Kingdom Department for International Development (DFID), 1999, Website: *www.unesco.org/csi/pub/source/rs.htm*, accessed 27 June, 2006.

Greidanus H. (2008), "Satellite Imaging for Maritime Surveillance of the European Seas", in Vittorio Barale and Martin Gade (2008), *Remote Sensing of the European Seas*, Springer, The Netherlands.

GTOS (2008), *Terrestrial Essential Climate Variables for Assessment, Mitigation and Adaptation*, GTOS-52, Food and Agriculture Organization of the United Nations, Rome, January.

Haimberger, L. (2007), "Homogenization of Radiosonde Temperature Time Series Using Innovation Statistics", *Journal of Climate*, Vol. 20, Issue 7, April.

Hay, S. (2006), "Marine Ecology: Gelatinous Bells May Ring Change in Marine Ecosystems", *Current Biology*, Vol. 16, Issue 17, pp. 679-682.

Hirsch, Robert M. (2004), "US Stream Flow Measurement and Data Dissemination Improve", *EOS*, Volume 85, No. 20, 18 May.

Høye, Gudrun K. *et al.* (2008), "Space-based AIS for Global Maritime Traffic Monitoring", *Acta Astronautica*, 62, pp. 240-245, January-February.

IMO (International Maritime Organization) (2006a), *International Shipping and World Trade: Facts and Figures*.

IMO (2006b), *Measures To Enhance Maritime Security, Long Range Identification And Tracking Of Ships*. MSC81/WP.5/Add.1, 18 May.

IOC/UNESCO (Intergovernmental Oceanographic Commission of UNESCO) (2005), *Proceedings of the Symposium on New Space Services for Maritime Users: The Impact of Satellite Technology on Maritime Legislation*, UNESCO, Paris, 21-23 February.

IOC/UNESCO (2007a), *IOC Strategic Plan for Oceanographic Data and Information Management (2008-2011)*, IOC Manuals and Guides No. 49, Intergovernmental Oceanographic Commission of UNESCO, Paris, 22 October.

IOC/UNESCO (2007b), "UNESCO and Inmarsat Sign Agreement to Improve Tsunami Warning System in Indian Ocean", UNESCO Press Release, No. 2007-162, UNESCO, Paris, 20 December.

ISU (International Space University) (2004), STREAM, Space Technologies for the Research of Effective Water Management, Student Team Project Final Report, International Space University, SSP Programme.

ITU (International Telecommunication Union) (2005), The Maritime Mobile and Maritime Mobile-Satellite Services, Geneva, 2005 Edition.

Jeannet, Pierre, Carl Bower and Bertrand Calpini (2008), Global Criteria for Tracing the Improvements of Radiosondes over the Last Decades, Report No. 95, World Meteorological Organisation, Geneva.

Jelenak, Zorana and Paul Chang (2008), NOAA QuikSCAT Follow-On Mission: User Impact Study Report, NOAA, Washington DC, 19 February.

Kallio K. et al. (2002), "Lake water quality classification with airborne hyperspectral spectrometer and simulated MERIS data", Remote Sensing of Environment, 79, pp. 51-59.

Kanellos, Michael (2006), "Satellites Used to Track World's Water Supply", CNET News.com, 12 December.

Kokkinos, M. et al. (2007), Broadband Satellite Networks and Triple Play Applications, IEEE 18th International Symposium on Personal, Indoor and Mobile Radio Communications (PIMRC 2007), Athens, Greece, 3-7 Sept.

Kongsberg Satellite Services (2008), Website: www.ksat.no, accessed March.

Kramer, Herbert (2002), Observation of the Earth and Its Environment, Survey of Missions and Sensors, 4th edition, Springer Verlag, Berlin.

Maillard, Catherine et al. (2007), "SeaDataNet: Development of a Pan-European Infrastructure for Ocean and Marine Data Management", SeaDataNet Consortium, 18-21 June.

NASA Goddard (2008), "Satellites Illuminate Pollution's Influence on Clouds", NASA Goddard Space Flight Press Release, 27 May.

NASA JPL (NASA Jet Propulsion Laboratory) (1996), "NSCAT First Image: First Wind Data from Scatterometer Captures Pacific Typhoons", NASA JPL Press Release, 3 October.

National Center for Atmospheric Research (2006), Innovative Satellite System Proves Its Worth with Better Weather Forecasts, Climate Data, December.

NOAA (National Oceanic and Atmospheric Administration) (2001), "NOAA'S Models and Satellites Help Chesapeake Bay Swimmers Avoid Jellyfish", NOAA Press Release, 2001-R307, 19 July.

NOAA (2008), NOAA's Website on Mapping Sea Nettles in the Chesapeake Bay, http://155.206.18.162/seanettles, accessed May.

NRC (National Research Council) (2006), Assessment of the Benefits of Extending the Tropical Rainfall Measuring Mission: A Perspective from the Research and Operations Communities, Interim Report, Committee on the Future of the Tropical Rainfall Measuring Mission.

NRC (2008), Earth Observations from Space: The First 50 Years of Scientific Achievements, Committee on Scientific Accomplishments of Earth Observations from Space, National Academies of Sciences, Washington DC, January.

OECD (2006), Ministerial Conference on International Transport Security, Tokyo, 12-13 January 2006, Summary of Findings, CEMT/CS(2006)27, European Conference of Ministers of Transport, Committee of Deputies, 22 March.

OECD (2007), *The Space Economy at a Glance*, OECD, Paris.

Payne, John F. *et al.* (2006), *Remote Sensing of the Bering Glacier Region*, 9th Bi-Annual Circumpolar Remote Sensing Symposium, 15-18 May.

Plant, W.J., W.C. Keller, and K. Hayes (2005), "Measurement of river surface currents with coherent microwave systems", *Geoscience and Remote Sensing*, Volume 43, Issue 6, June.

Robinson, Ian Stuart (2004), *Measuring the Oceans from Space: The principles and methods of satellite oceanography*, Springer Praxis, New York.

Roquet, Hervé (2006), "MERSEA Satellite SST Activities and Products", *Mercator Ocean Quarterly Newsletter,* Vol. 22, July.

Serra-Sogas, Norma *et al.* (2008), "Visualization Of Spatial Patterns And Temporal Trends For Aerial Surveillance Of Illegal Oil Discharges In Western Canadian Marine Waters", *Marine Pollution Bulletin*, Vol. 56, Issue 5, May.

ShipLoc (2008), Company's website, *www.shiploc.com*, accessed 12 April.

Spector, Laura (2007), *R.I.P. TOMS: NASA Ozone Instrument Laid to Rest after Three Decades*, *NASA Goddard Space Flight Center* (Website: *www.nasa.gov/centers/goddard/news/topstory/2007/toms_end.html*, 15 August.

Statheros, Thomas, Gareth Howells and Klaus McDonald-Maier (2008), "Autonomous Ship Collision Avoidance Navigation Concepts, Technologies and Techniques", *Journal of Navigation,* Vol. 61, Issue 01, January, pp. 129-142.

Tunaley, J.K.E. (2004), "Algorithms for Ship Detection and Tracking Using Satellite Imagery", *Proceedings of Geoscience and Remote Sensing Symposium 2004*, IGARSS, IEEE International, Vol. 3, 20-24 Sept.

UNESCO (2008), "Ocean Observing Flotilla Hits 3000 Mark", *Natural Sciences*, Quarterly Newsletter, Vol. 6, No. 1, January-March.

US Coast Guards (2008), *How to Conduct a GMDSS Inspection*, Annex E of Federal Communications Commission Docket No CI- 95-55, adopted 20 April 1998, by Report and Order FCC No. 98-75.

Vaneli-Corali, A. *et al.* (2007), "Satellite Communications: Research Trends and Open Issues", *International Workshop on Satellite and Space Communications* (IWSSC '07), Salzburg, Austria, 13-14 Sept.

Vitaly, Alexandrov *et al.* (2008), "Detection of Arctic Icebergs on the Base of Satellite SAR", Proceedings of the SEASAR2008 Worshkop, Advances in SAR Oceanography from ENVISAT and ERS missions, ESA ESRIN, Frascati, Italy, 21-25 January.

Wahl T., and Skøelv Å, Pedersen J P, Seljelv L G, Andersen J H, Follum O A, Dahle Strøm G, Anderssen T, Bern T I, Espedal H, Hamnes H, Solberg R (1996), "Radar satellites: A new tool for pollution monitoring". *Coastal Management*, Vol. 24, No. 1, Jan 1996, pp 61-71.

Winther, Jan-Gunnar *et al.* (2005), "Remote Sensing of Glaciers and Ice Sheets", *Remote Sensing in Northern Hydrology: Measuring Environmental Change*, Geophysical Monograph Series, American Geophysical Union, pp. 39-62.

WMO (World Meteorological Organisation) (2008), *Space-Based GOS Vision for 2025*, WMO Secretariat, Consultative Meetings on High-Level Policy on Satellite Matters, Eighth Session, New Orleans, Louisiana, 15-16 January.

ISBN 978-92-64-05413-4
Space Technologies and Climate Change
Implications for Water Management, Marine Resources
and Maritime Transport
© OECD 2008

Chapter 5

Outlook for Space Technologies

The demands for climate information, resource sustainability, transport efficiency, etc. are growing quickly. Can the space sector deliver the research, technology and management to keep pace with so much rapid change, now and in the future? The many successes of space assets may obscure some of the technical limitations as well as governance related restrictions, and so run the risk of overselling what satellites can and will do. This chapter shows that more applications are in the pipeline or are already coming on stream; if gaps are to be remedied, substantial investments in earth observation and meteorological satellite systems will be required over the next ten years. In addition, replacement and network expansion investments will be required thereafter to address the mounting challenges to 2025 and beyond.

Remedying technical limitations

Whether for earth observation, commercial communications or navigation, the development of new systems is accelerating. There have never been so many "connections and eyes in the sky" as there are today, providing signals and data useful for climate research and monitoring. And more are planned. But efforts are needed for continuous observations and to build data links, as well as to integrate the data into terrestrial information systems.

Earth observation

Earth observation systems already provide key information for a large number of applications. Some gaps remain however, which could be addressed at least in part by technical developments.

Satellite remote sensing sometimes suffers from poor temporal resolution. Some water cycle phenomena occur quickly and need to be followed in near-real-time; examples include flooding, storm cloud development and the spread of water pollution (*e.g.* oil slicks). Geostationary satellites can monitor areas continuously but not with the required spatial resolution, while satellites in other lower orbits (*e.g.* polar orbits) do not have the necessary revisit frequency. Meeting both the temporal and spatial requirements for monitoring "fast" phenomena will increasingly call for constellations of satellites in low earth orbit. As an example, the Italian Cosmo Skymed radar satellite constellation (COnstellation of small Satellites for Mediterranean basin Observation), with its planned four satellites (the last one to be launched in mid-2009), will be able to revisit any point in Europe every 4.5 to 12 hours. With regard to responsiveness, it will take 12 hours to respond to an urgent request to collect data over a specific area (44 hours in a routine mode) (Coletta *et al.*, 2008).

The number of spectral bands being monitored is still limited. Through observation of a range of bands in the electromagnetic spectrum, various characteristics of the climate and water cycle can be monitored. However, current satellite sensors do not fully cover the range of spectrum bands required for effectively monitoring characteristics like salinity and water quality.

Satellite remote sensing delivers sometimes poor spatial resolution. There is an inherent trade-off in current satellite technology in that sensors with multispectral or hyperspectral capabilities that are needed to accurately determine some characteristics (*e.g.* water temperature) have poor spatial resolution. This gap in technology has to be remedied through integration with airborne and ground-based monitoring systems that have both multiple bands and sub-metre spatial resolution.

Calibration of space remote sensing data is difficult. Detailed *in situ* measurements can be difficult to gather for large parts of the earth, especially oceans: it is very expensive to collect such "ground truth" measurements with sufficient spatial resolution in remote areas.

The telecommunications sector too could see improved capabilities via technical advances. The increased onboard computational analysis capacities of satellites ("intelligent satellites in the sky") should reduce the bandwidth requirements for data download. This allows for the use of more sampling channels on the sensors. For example, if several new spectral analysis algorithms are applied immediately to outputs from a hyperspectral sensor on board a satellite, the data transmitted back to earth already benefits from basic treatment and may be more readily usable. With regard to ground segments, the continuing upgrades of computing power and networking capacities are steadily reducing the delay between acquisition and delivery of processed ocean and more general water data products to users. This enhances the usefulness of real-time data operational activities.

Navigation and localisation

The outlook for ship navigation, detection and information using satellite technologies is promising. It builds on two main trends: the further deployment of navigation systems and future possible constellations of Automatic Identification System (AIS) satellites.

There are six current and prospective providers of Global Navigation Satellite Systems (GNSS) and related navigation signal augmentations as of Spring 2008: the United States, the Russian Federation, the European Union, China, Japan, and India (ICG, 2008).

- GPS is currently the only fully operational navigation satellite system with a global coverage. Its modernisation is under way: new satellites are being launched (modernised GPS "Block IIR") and new signal augmentation capabilities implemented in the GPS Wide-area Augmentation System (WAAS). A new GPS generation will be launched (GPS III) as of 2014, with better accuracy and availability for both civilian and military users.

- The Russian Federation is also modernising its Glonass system. The constellation is currently operating in a degraded mode, with only 12 satellites fully operational (Russian Space Agency, 2008). The regularly launched Glonass-M satellites have better signal characteristics as well as a longer lifespan (seven or eight years instead of three years) and the transition to a third generation of Glonass-K satellite with a lifespan of

10 years is under way. The objective is to have a fully working constellation by 2010. New joint GPS-Glonass signal receivers are also being developed, particularly for Russian maritime users.

- In Europe, Galileo went through a difficult phase with the failure of the public-private partnership approach, but has been back on track since early 2008. Full operational capability of the 24-satellite constellation could be reached by 2013. Already the European Geostationary Navigation Overlay Service (EGNOS) is providing GPS augmentation signals over Europe.

- China is developing the Beidou/Compass navigation system; its architecture is different from the GPS, Glonass and Galileo systems, all of which use medium earth orbit (MEO) satellites. Beidou is designed to feature five satellites in geostationary orbit and 30 satellites in medium earth orbit. Demonstrations are taking place and by 2010 China expects to have an operational system – not yet global, but covering large parts of Asia.

- India is developing its own GPS space-based augmentation signal system using ground-based stations and an Inmarsat satellite in geostationary orbit. The system, called GPS Aided Geo Augmented Navigation (GAGAN), is designed to send more precise signals over Indian airspace. It should be operational by 2010. An independent regional navigation system is also under development, with the objective of covering 1 500 kilometres around India using signals from three geostationary satellites and four geosynchronous satellites. That system could be operational by 2011.

- Other systems include Japan's GPS augmentation system, the Quasi Zenith Satellite System (QZSS) which is planned to be launched before March 2010. The QZSS system will ultimately consist of three satellites.

Discussions on the compatibility and interoperability of these systems are due to continue, particularly through the recently created International Committee on Global Navigation Satellite Systems at the United Nations (ICG, 2008). Technological progress should generally provide more accurate and reliable signals in the future, while the miniaturisation of chipset and software will lower even further the costs of integrating navigation signals into a wide range of equipment. However, other ground-based systems (*e.g.* mobile phone networks) may require or compete for integration with such systems, and augmentation systems need further improvement to allow safe GPS positioning for some sensitive applications (such as aviation) (Table 5.1).

In addition to those ongoing developments in satellite navigation, research continues on specific constellations targeted at ship detection and information. Many countries already use coastal radar chains, vessel monitoring systems for fishing fleets, and coastal Automatic Identification System (AIS) chains. In the future, a constellation of satellites will be needed for global maritime traffic monitoring, as suggested by Høye *et al.* (2008). This would require posting AIS

Table 5.1. **Satellite navigation augmentation systems (ground- and space-based)**

Ground-based augmentation systems (GBAS)	**Differential global positioning systems (DGPS)** DGPS are ground-based systems (stations and antennas) that enhance the accuracy of GPS signals for maritime and terrestrial users by providing corrections to the user's receiver in real time. Those correction systems may be: – Public (*e.g.* the US Coast Guard's DGPS), allowing the user to get the corrected signal with any DGPS receiver. – Or private (*e.g.* Omnistar); in that case, users need to subscribe to the DGPS provider to receive the specific enhanced signal. The DGPS signals can have an accuracy of 1 to 5 metres depending on the quality and price of the GPS receiver, and even be sub-metric in the case of some private enhanced signal providers. DGPS have already been adopted globally as an international maritime standard established by the 1994 International Telecommunication Union. Over 40 countries are implementing DGPS services. **Local area augmentation systems (LAAS)** LAAS, in general based locally at airports, provide increased position accuracy by sending differential correction signals to aircrafts.
Space-based augmentation systems (SBAS)	SBAS are based on a compatible navigation payload aboard one or more geostationary satellites (36 000-kilometre orbit) over a specific region, supported by the necessary ground segment and uplink earth stations. Signals from GPS or Glonass can be rectified and enhanced by cross-checking with the geostationary satellites' payload [*e.g.* wide area augmentation systems (WAAS) in the United States and EGNOS in Europe].
Aircraft-based augmentation systems (ABAS)	ABAS are installed in planes to provide local augmentation systems. They include: – Remote aircraft integrity monitoring (RAIM). – Future air navigation system (FANS). – Aircraft autonomous integrity monitoring (AAIM).

receivers in space, as a dedicated micro- or nano-satellite mission or as an additional instrument on larger satellites. Coastal authorities can normally ask ships outside coastal areas to transmit their information via satellite (Inmarsat), via the long-range option of the AIS system. But this reporting option is not mandatory. A constellation of AIS satellites could provide updated ship information as often as once per hour in areas with low ship density (Table 5.2).

Table 5.2. **Possible system performance for a space-based AIS system**

	Ship information update rate	Number of ships	Ship detection probability
Single satellite	1 per satellite pass	900	> 99%
Regional constellation	1 per hour	1 300	> 99%
Global constellation	1 per day	2 100	> 99%

Source: Høye et al., 2008.

One problem that arises when an area is dense with ships is that all their AIS signals are blended, and so identifying specific vessels based on their respective signals from space is difficult if not impossible (Cauzac and Lebras, 2008). In European waters for example, an AIS sensor at 1 000-kilometre altitude with a field of view to the horizon (a 3 630 nautical mile-wide swath) may see up to 6 200 ships that carry AIS simultaneously, and cannot discriminate among them (Høye *et al.*, 2008).

This is also a problem faced by operators when they try to combine different ground-based surveillance systems. In Spain there was a deliberate choice not to merge data from two different surveillance systems that cover the same area but with different resolutions [*i.e.* the Spanish Integrated System of External Vigilance (SIVE) and VTS systems] in order to avoid blurry maps with too many ships of different sizes (EC JRC, 2008). There are, however, plans to integrate AIS data in SIVE.

A possible new standard for LRIT based on the existing AIS system is suggested to solve this problem. Such a system could give global coverage and updated ship information 1-4 times per hour. Another technical alternative consists of enabling the AIS satellites to do more onboard processing, by demodulating and extracting the ships' signals before downloading them to ground stations. That alternative is still very much work in progress, but Norway is already envisaging a demonstrator satellite launch in the 2009-10 time frame (Milsom, 2008).

Communications

Communications satellites form a special type of infrastructure, comparable to "information highways in the sky" that are operated mainly by commercial providers. This area of communications already represents a robust market for operators, but challenges remain to improve services and respond to new market demands.

The main technical challenge is to develop new generations of communications satellites offering on board processing and spot beams. Adoption of technical standards is also necessary, to facilitate scalability and further reduce costs (notably that of terminals). Current telecommunications satellites are mostly designed as simple relays, using so-called "bent-pipe transponders" that transmit the signals they receive with a minimum of processing; they thus load most of the complexity of the telecommunications process into the ground segment.

Current work in onboard processing paves the way for more complex satellites, with multiple spot beams that are able to operate as "routers in the sky"; these become nodes in the global communications network, enabling reuse of scarce frequency resources. New systems have already come online, especially to develop broadband solutions. This progress is currently leading to a new generation of commercial satellite constellations for telephony and broadband; these benefit from smaller-size antennas and terminals (laptop-like in some cases) and better performance.

Another trend is towards "integrated applications" (telecom- and location-based). Location-based technology has matured to the point where it can now be deployed in support of a wide range of business applications. Moreover, the so-called "GYM" club (Google, Yahoo and Microsoft) has raised consumer demand and expectations to new levels. Future killer applications will require the development of very high broadband networks (OECD, 2005). The near future will see moves from triple play (voice, video, data from the Internet) and quadruple play (+ mobile telephony) to quintuple play and beyond (geo-localisation, etc.). Indeed, the traditional separation between telecommunications and content providers is already blurring.

This expansion of the overall space infrastructure, although welcomed by existing users, is no guarantee that adequate services will follow. Key challenges remain concerning the actual development of the required systems and the sustainability of communications, navigation and, as the following section demonstrates, earth observation.

The future of earth observation for climate monitoring

Earth observation provides important information, as seen in the previous chapters. However, there are still gaps in coverage that sometimes limit the adequacy of the sensors available. There are also governance issues relating to sustainability of the systems. The outlook for setting up an improved international global observing system (GOS) by 2015 and 2025 is favourable. But this international effort will require sustained investments.

Tackling governance issues for the sustainability of systems

In parallel with the growing need for reliable climate data gathered from space-based earth observation, there has emerged a more systematic approach to the use of space-based capabilities nationally and internationally. A number of co-ordinating bodies have been involved (Table 5.3). However there are limitations, linked more often to governance issues than solely to technical problems.

As numerous organisations are currently measuring climate parameters for a variety of purposes, different measurement protocols are used. This results in a lack of homogeneity in the data (in space and time), which in turn limits use of the data for many applications and constrains the scientific capacity to monitor and assess climate change.

There is also a certain "fragility" of observing systems – especially those functioning on research funds, as with the ocean observations (CEOS, 2007). Transitioning observing systems from research to operation is a serious governance problem in many countries that is not yet resolved. On the level of space programme management, the need for continuity of measurement of a particular climate variable does not necessarily imply that new types of

Table 5.3. **Examples of international organisations promoting co-operation in climate-related observations**

Organisation	Description
Intergovernmental Panel on Climate Change (IPCC)	IPCC was established by WMO and UNEP to assess scientific, technical and socio-economic information relevant for understanding climate change, its potential impacts and options for adaptation and mitigation.
World Climate Research Programme (WCRP)	Established in 1980, WCRP's objectives are to develop sufficient fundamental scientific understanding of the climate system and its processes to determine the extent to which climate can be predicted and is subject to human influence. The Global Energy and Water Cycle Experiment (GEWEX) is an international programme to observe, understand and model the hydrological cycle and energy fluxes in the atmosphere, at land surface and in the upper oceans. Its modelling activities contributed to the Fourth Assessment Report of the IPCC, published in 2007.
Global Climate Observing System (GCOS)	Established in 1992 to ensure that the observations and information needed to address climate-related issues are obtained and made available to all potential users. GCOS works closely with many bodies, including the World Meteorological Organization, to co-ordinate and develop further the international Global Observing System (GOS).
Committee on Earth Observation Satellites (CEOS)	CEOS is a major international forum with 25 members, most of which are space agencies. Its role is to co-ordinate earth observation satellite programmes.
Group on Earth Observations (GEO)	GEO is leading a worldwide effort to co-ordinate a Global Earth Observation System of Systems (GEOSS) over the next ten years.

instruments have to be developed. This is a dilemma faced by many R&D space agencies that are asked by the climate community to continue observational capabilities. If they do so they must include possible overlap satellite missions for the preservation of satellite climate data records (Figure 5.1).

Figure 5.1. **Likely gaps in satellite altimetry records by 2015**

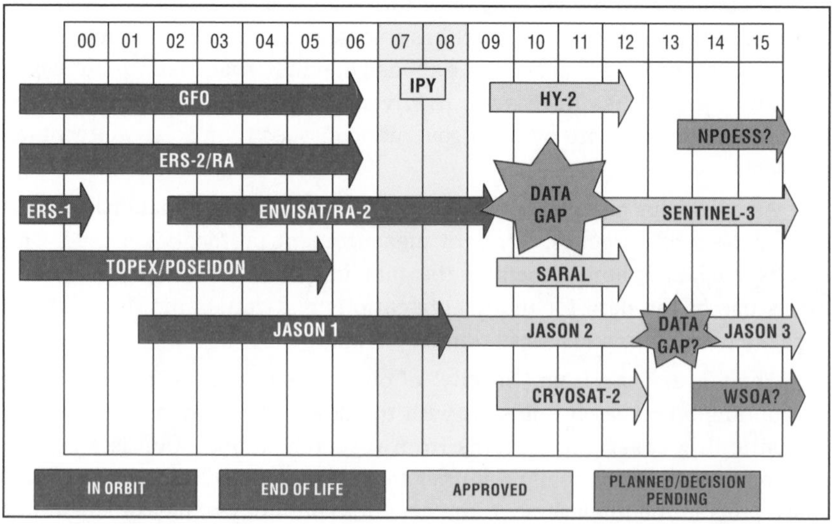

Source: Adapted and updated from CEOS, 2007.

Another governance issue concerns access to the data. Databases containing space remote sensing data are often not complete, nor easily accessible, nor sufficiently quality controlled. Large observational gaps in area coverage and coverage period are the norm, according to international science groups (GCOS, 2005). There are neither general standards nor databases, and the multitude of formats renders access to the data difficult, let alone its manipulation. It is estimated that by August 2007, 59% of the oceans' *in situ* networks were complete (Alverson, 2007). But they are far from being used in a co-ordinated way with the numerous new satellite optical and radar data sources now or soon available (*e.g.* Cosmos Skymed and TerraSAR-X data).

Towards an international global observing system (GOS) in 2015 and 2025

These shortcomings are progressively being addressed as plans evolve for the future international global observing system. GOS by definition includes both ground-based systems (*e.g.* buoys at sea, ships, weather radars) and space components.

Today the global observing system already comprises three complementary space-based constellations: operational meteorological polar orbiting satellites (*e.g.* MetOp); operational meteorological geostationary satellites (*e.g.* GOES satellites, MTSAT, ELEKRO-L, INSAT, FY-2/4); and environmental research and development satellites (*e.g.* ENVISAT, Quickscat). These were developed by different countries such as the United States, Europe, China, India, Japan, Korea and the Russian Federation. As of January 2008, nine geostationary satellites and seven low earth orbit satellites were declared in operational mode.[1] Nineteen additional R&D satellites were also contributing to the GOS.[2]

The focus until recently has been mainly on operational meteorology via the World Weather Watch co-ordination programme. This principal programme of the World Meteorological Organisation, set up in 1963, combines independent observing systems, telecommunication facilities, and data processing and forecasting centres (WMO, 2008). Based on the current plans from space agencies and meteorological organisations, many climate variables should in the near future be observed over large parts of the world with long-term continuity constraints and on the level of the World Weather Watch programme, and not only by potential R&D missions.

A co-ordinated Global Observing System could by 2015 include a more diversified fleet of independent satellites, covering more weather and climate variables as agreed by a special WMO Extraordinary Session in 2002:

- At least six operational geostationary satellites – all with multispectral imager (Visible Infrared IR/VIS); some with hyperspectral sounder (Infrared IR).

- Four operational low earth orbit (LEO) satellites optimally spaced in time – all with multispectral imager (MW/IR/VIS/UV), all with microwave (MW) sounder, three with hyperspectral sounder, two with altimeter, three with conical scan MW or scatterometer.

- A constellation of small R&D satellites for radio occultation – in low earth orbit with wind lidar, in LEO with active and passive microwave precipitation instruments, and in LEO and geostationary with advanced hyperspectral capabilities.

- And improved intercalibration and operational continuity (WMO, 2002).

One relatively new element is the possibility of having as early as 2015 improved thematic constellations (*e.g.* hydrology, oceanic missions). The A-Train constellation, which enjoyed some success, was mentioned in Chapter 4; it includes independent R&D satellites that, when used together, provide key climate-related information from US and French satellite sensors in near-real-time.

Looking further into the future, the GOS is developing a vision for 2025: with inputs from a large number of organisations, it aims to support improved operational weather forecasting, climate monitoring, oceanography, atmospheric composition, hydrology, and weather and climate research. Preliminary work has already identified a number of parameters (Eyre, 2008; see Table 5.4):

- Increased collaboration between space agencies is necessary, so that user requirements for observations are met in the most cost-effective manner, and that system reliability is assured through arrangements for mutual back-up.

- The observational capability demonstrated on R&D satellites needs to be progressively transferred to operational platforms, to assure the reliability and sustainability of measurements.

- R&D satellites should continue to play an important role in the GOS. Although they cannot guarantee continuity of observations, they offer important contributions beyond the current means of operational systems. Partnerships will need to be developed continuously between agencies to extend the lifetime of operational and other R&D satellites to the maximum useful period.

- Increasingly, user requirements will have to be met through constellations of satellites, often involving collaboration between space agencies. Expected satellite constellations missions could include altimetry, precipitation, radio occultation, atmospheric composition and earth radiation budget.

- Availability and timeliness will have to be improved through operational co-operation among agencies.

Table 5.4. **Possible components of a space-based global observing system by 2025**

Platform	Instruments	Observed variables
Operational geostationary (GEO) satellites At least 6, nearly equally spaced	Visible infrared imagers (VIS/IR)	Cloud amount, type, top height/temperature; wind (through tracking cloud and water vapour features); sea/land surface temperature; precipitation; aerosols; snow cover; vegetation cover; atmospheric stability; fires
	Infrared hyperspectral sounders	Atmospheric temperature, humidity; wind (through tracking cloud and water vapour features); sea/land surface temperature; cloud amount and top height/temperature; atmospheric composition
Operational polar-orbiting sun-synchronous satellites 3 orbital planes	Infrared hyperspectral sounders	Atmospheric temperature, humidity and wind; sea/land surface temperature; cloud amount, water content and top height/temperature; atmospheric composition
	Microwave (MW) sounders	
	Visible infrared imagers	Cloud amount, type, top height/temperature; wind (high latitudes, through tracking cloud and water vapour features); sea/land surface temperature; precipitation; aerosols; snow and ice cover; vegetation cover; atmospheric stability
Additional operational missions in appropriate orbits	Microwave imagers, at least 3, some polarimetric	Sea ice; total column water vapour; precipitation; sea surface wind speed (and direction); precipitation; cloud liquid water
	Scatterometers – at least 2	Sea surface wind speed and direction; sea ice; soil moisture
	Radio occultation constellation – at least 6	Atmospheric temperature and humidity; ionospheric electron density
	Altimeter constellation	Ocean surface topography; sea level; ocean wave height; lake levels
	Infrared dual-angle view imager	Sea surface temperature (of climate monitoring quality); aerosols; cloud properties
	Narrow-band visible near infrared imager	Ocean colour; vegetation (including burned areas); aerosols; cloud properties
	High-resolution visible infrared imagers – constellation	Land-surface imaging for land use and vegetation
	Active and passive microwave instruments – constellation	Precipitation
	Broadband visible infrared radiometer + total solar irradiance sensor – at least 1	Earth radiation budget (supported by imagers and sounders on polar-orbiting and geostationary satellites)
	Atmospheric composition instruments – constellation	Ozone; other atmospheric chemical species; aerosols – for greenhouse gas monitoring, ozone/UV monitoring, air quality monitoring
	Synthetic aperture radar	Wave heights, directions and spectra; oil spills; floods; other hazards; earthquake and faults monitoring; sea ice leads; damage assessment; ice shelf and icebergs

Table 5.4. **Possible components of a space-based global observing system by 2025** *(cont.)*

Platform	Instruments	Observed variables
Capability on R&D satellites and operational pathfinders	Doppler wind lidar on low earth orbit (LEO)	Wind; aerosol; cloud-top height (and base)
	Low-frequency microwave radiometer on LEO	Ocean surface salinity; soil moisture
	Microwave imager/sounder on GEO	Precipitation; cloud water/ice; atmospheric humidity and temperature
	Lightning imager on GEO	Lightning, location of intense convection
	Visible infrared imagers, imagers on satellites in highly elliptical orbit (HEO)	Winds and clouds at high latitudes; sea ice
	Gravimetric sensors	Water volume in lakes, rivers, ground, etc.
	Others	Three-dimensional cloud water/ice fields; sea and land ice topography; flood monitoring

Source: Based on Eyre, 2008.

Required investment in earth observation

In view of the rapidly increasing demand for data on climate change, weather conditions, monitoring resources and ocean movement tracking, it is clear that substantial investments in earth observation and meteorological satellite systems will be needed over the coming years.

Learning from past investments

Trying to estimate how much has been invested in the earth observation sector over the years is a daunting task, especially from an international perspective. Literally hundreds of billions of US dollars have been invested in R&D since the dawn of the space age to get to the systems we have in orbit today. It is however useful to have an idea of the past investment that led to the existing space-based infrastructure – as imperfect as the estimates may be in some cases – in order to make some assumptions about the investments to come. An attempt is therefore made here to assess conservatively the value of earth observation and meteorological space assets already in orbit, and to estimate the annual cost of maintaining and upgrading them.

The first step is to try to establish how many relevant satellites are currently in orbit. There is no available real-time, independent, comprehensive, global registry.[3] Estimates of the number of active satellites in orbit vary, but it seems reasonable to suggest that they total around 900. Many of these are military/strategic, and most are dedicated to telecommunications. What is of interest here are the satellites that may have a bearing, even if limited, on climate change management. These include earth observation satellites of different types (*i.e.* land imaging and ocean observation satellites), some scientific satellites, meteorological satellites, and a few declared dual-use satellites. Many of these satellites are, of course, multipurpose, as seen in Chapter 4; very diverse applications can be derived from data they provide.

The OECD list established for the purposes of this exercise contains 100 civilian earth observation satellites (including 20 meteorological satellites) active at the end of 2006 – the figure is very close to the most recent American Society of Photogrammetry and Remote Sensing (ASPRS) list of 102 satellites (with resolutions greater than 39 metres), and the larger list compiled by the Union of Concerned Scientists – 107 satellites.[4] The actual figure of earth observation satellites is thought to be larger, since not all relevant Russian and Chinese satellites have been identified and a few active satellites from other countries are also thought to be missing from the list. The majority of the identified remote sensing satellites are deployed to low earth orbits (around 80 satellites), while a smaller quantity of satellites (around 20) are deployed to geostationary orbits, primarily for global weather forecasting.

Based on the OECD list – and using publicly available cost estimates and a GDP deflator to calculate costs at current USD rates (Box 5.1) – early conservative estimates place the value of the total stock of active earth observation satellites relevant to climate and water resources management at around USD 20 billion (current). The 20 civilian meteorological satellites in orbit at the time of the study and identified on the OECD list (all operational, although some are in standby mode and a few are multifunction, *i.e.* telecom) have an estimated value of some USD 8.75 billion. The value of the other 80 earth observation satellites recorded for the purposes of the study is thought to be approximately USD 11 billion.

Annual investments – considered to be maintenance, replacement and expansion of the overall earth observation space-based infrastructure system – vary depending on the rhythm of launches, the active lifespan of satellites, their size, function and cost, etc. Taking into account those caveats in the variables, annual investments are estimated for 2006, 2005 and 2004 in Table 5.5.

On the basis of the OECD list of satellites in orbit, the value of earth observation satellites launched in 2006 – a particularly active year – amounted to USD 3.2 billion or around 15% of total space-based earth observation and meteorological infrastructure. The figure for 2005 was approximately USD 1.1 billion or 6% of total in-orbit assets, and in 2004 around USD 1.6 billion or 10% of total active satellite infrastructure. Since then, new earth observation satellites were launched in 2007, while a record number are expected in 2008 – 17 satellites in that year alone (WMO, 2007). Those estimates for earth observation will be compared in the next section with other infrastructure costs.

Box 5.1. **Methodology used to assess the 2006 stock value of active earth observation satellites**

Conservative public estimates have been used to value the costs of active civilian earth observation satellites launched from 1990 to 2006. A total cost for each given mission was estimated, using public data (e.g. agency or industry reports, press releases); these may not always clearly discriminate between launch costs, satellite costs and operating costs. The figures used therefore remain largely conservative: a) many satellites have been operating in orbit longer than their design lifetime (i.e. inducing higher operating costs than the ones anticipated in some of the available estimates); b) the found values may underestimate the overall costs of the missions (especially when estimates are not official, as in the case of China or Russia); and c) delays in launching or possible costs overruns in developing ground-based infrastructures (for operations) may not appear in the total value of the systems. Finally, a satellite mission is the result of years-long research and development (R&D) investments that are usually not translated into the systems' available costs estimates.

Based on the figures found, a methodology was developed to put past values of satellite missions launched into present values. Although there were a number of specific variables that could have been used to do so (e.g. producer price index, consumer price index, etc.), the "Deflator for GDP at Market Prices for the OECD Region" was used for a number of reasons. Firstly, there was no particular variable that seemed uniquely geared towards representing the changing value of producing, launching, servicing satellites over this period, making GDP a good variable for general price changes in the economy. Secondly, many variables that were available often lacked values for key international players or for certain periods. Thirdly, a large number of satellite launches over this period were from OECD countries. The particular calculation that was performed was an "inflation" of past values into present value terms. In most cases, the value of an item into the prices of a particular base year. Here, the opposite occurred. For example, if it cost USD 100 to produce an item in 2005 and USD 110 to produce it in 2006, the 10% change in prices that would usually be used as a "deflator" to put 2006 values into 2005 values was used as an "inflator" to increase the value of 2005 into 2006 terms. Thus both outputs would now be valued at USD 110. This method was used to move all items back to 1990 in 2006 value terms.

Table 5.5. **Estimated annual investments (maintenance, replacement, expansion) in earth observation (2006, 2005, 2004)**

	Annual investments (in billion USD and as % of total in-orbit assets at end-2006)	
2006	3.2 billion USD	15
2005	1.1 billion USD	6
2004	1.6 billion USD	10

Future investments

The earth observation sector is experiencing many changes; new actors are undertaking their own space programmes (OECD, 2005). The development of strong earth observation programmes in Asia (*e.g.* India, China), South America and Africa is an encouraging factor, as more institutional actors than ever before are involved in developing systems. The multiplication of commercial initiatives and systems from which climate data can also be derived is another positive sign. But despite a more evenly distributed workload thanks to international co-operation, developing the necessary systems identified above in the GOS 2015 and 2025 visions will come at a cost.

In the United States, for example, a recent report by the US National Academy of Sciences (2007) has estimated conservatively that investments in earth observation systems of around USD 7.5 billion over the next 15 years are necessary for the information needs of US policy makers, scientists and operational users. This comes in addition to investments in meteorological satellites (around USD 12 billion for the next generation fleet, which will combine DOD and NOAA satellites).

In the case of Europe, national and European-level investments are also ongoing for earth observation. As an example, ESA has three main programmes under way: the Explorer missions (designed for research purposes), the Sentinels (operational satellites focusing on the continuity and improvements of existing services exploiting earth observation data) and collaboration with Eumetsat in meteorology. ESA's Explorer missions each have costs ranging from around EUR 100 million to EUR 400 million per mission, with investment already above EUR 1.5 billion as shown in Table 5.7. The initial Sentinel investments, shared by the European Space Agency's member states and the European Commission, should be around approximately EUR 2.3 billion; that covers R&D, development, ground segment and data access, although further funds will be required to sustain operational aspects.[5] Finally, developments are ongoing in the European meteorological sector. The Eumetsat MetOp Polar System, a series of three satellites (and associated ground segment) to be launched sequentially to deliver data to monitor climate and improve weather forecasting until at least 2020, is due to cost around EUR 2.4 billion (MetOp-A, the first satellite, was launched in 2006).[6]

Many countries are also pursuing earth observation programmes as national initiatives and in bilateral co-operation (*i.e.* France, Germany, Italy, the Russian Federation, India, and China to name a few). Based on the needs identified for renewal of existing systems and the development of new systems, a very conservative guess would point to worldwide investments in space-based earth observation of around USD 38-40 billion by 2020. These indicative estimates of earth observation and meteorological satellite

Table 5.6. **Missions recommended by the US National Research Council with estimated costs (January 2007)**

Name of mission	Main objectives	Orbit	Estim. costs (million USD)	Agency
PERIOD 2010-2013				
CLARREO	Solar and earth radiation characteristics for understanding climate forcing and response of the climate system	LEO, SSO	265	NOAA/NASA
GPSRO	High-accuracy, all-weather temperature, water vapour and electron density profiles for weather, climate and space weather	LEO	150	NOAA
SMAP	Soil moisture and freeze/thaw for weather and water cycle processes	LEO	300	NASA
ICESat-II	Ice sheet height changes for climate change diagnosis	LEO, Non-SSO	300	NASA
DESDynI	Surface and ice sheet deformation for understanding natural hazards and climate; vegetation structure for ecosystem health	LEO, SSO	700	NASA
		Subtotal:	**1 715**	
PERIOD 2013-2016				
XOVWM	Sea surface wind vectors for weather and ocean ecosystems	LEO, SSO	350	NOAA
HyspIRI	Land surface composition for agriculture and mineral characterisation; vegetation types for ecosystem health	LEO, SSO	300	NASA
ASCENDS	Day/night, all-latitude, all-season CO_2 column integrals for climate emissions	LEO, SSO	400	NASA
SWOT	Ocean, lake and river water levels for ocean and inland water dynamics	LEO, SSO	450	NASA
GEO-CAPE	Atmospheric gas columns for air quality forecasts; ocean colour for coastal ecosystem health and climate emissions	GEO	550	NASA
ACE	Aerosol and cloud profiles for climate and water cycle; ocean colour for open ocean biogeochemistry	LEO, SSO	800	NASA
		Subtotal:	**2 850**	
PERIOD 2016-2020				
LIST	Land surface topography for landslide hazards and water runoff	LEO, SSO	300	NASA
PATH	High-frequency, all-weather temperature and humidity soundings for weather forecasting and SST	GEO	450	NASA
GRACE-II	High temporal resolution gravity fields for tracking large-scale water movement	LEO, SSO	450	NASA
SCLP	Snow accumulation for fresh water availability	LEO, SSO	500	NASA
GACM	Ozone and related gases for intercontinental air quality and stratospheric ozone layer prediction	LEO, SSO	600	NASA
3D-Winds (demo)	Tropospheric winds for weather forecasting and pollution transport	LEO, SSO	650	NASA
		Subtotal:	**2 950**	
		Total:	**7 515**	

Source: NRC, 2007.

Table 5.7. **Planned ESA Explorers and Sentinels earth observation missions (2008-2013) (June 2008)**

	Name of mission	Main objectives	Estim. costs (million EUR	Planned launch date
PLANNED ESA EARTH EXPLORERS MISSIONS	GOCE (Gravity Field and Steady-State Ocean Explorer)[1]	Measurements of the earth's gravity field and modeling of the geoid to advance knowledge of ocean circulation, geodesy and surveying. Mission duration: about 20 months.	200	Sept. 2008
	SMOS (Soil Moisture and Ocean Salinity)[2]	Observation of soil moisture over the earth's landmasses and salinity over the oceans. Mission duration: 3 to 5 years.	200	2009
	ADM-Aeolus (Atmospheric Dynamics Mission)[1]	First space mission to measure wind profiles on a global scale, to improve accuracy of numerical weather forecasting and advance understanding of atmospheric dynamics and processes relevant to climate variability and climate modelling. Mission duration: at least 3 years.	300	2009
	CryoSat-2[2]	Measurements of the thickness of floating sea ice to detect seasonal to inter-annual variations, and also survey the surface of continental Antarctic and Greenland ice sheets. CryoSat-2 replaces CryoSat, which was lost due to a launch failure in October 2005. Mission duration: at least 3 years.	106 (136 for CryoSat-1)	2009
	Swarm[2]	Constellation of three satellites that will provide high-precision and high-resolution measurements of the strength and direction of the earth's magnetic field, to study atmospheric processes related to climate and weather, space weather and radiation hazards.	88 (contract signed in 2005)	2010
	EarthCARE (Earth Clouds, Aerosols and Radiation Explorer)[1]	Mission implemented in co-operation with JAXA to study the interactions between cloud, radiative and aerosol processes that play a role in climate regulation.	263 (contract signed May 2008)	2013
PLANNED ESA SENTINELS	SENTINEL-1	Sentinel-1 is an all-weather, day-and-night radar imaging satellite mission for land and ocean services.	229 (contract signed June 2007)	2011
	SENTINEL-2	Sentinel-2 is a high-resolution optical imaging mission for land services. The mission objective is systematic coverage of the earth's land surface (from −56° to +83° latitude) to produce cloud-free imagery typically every 15 to 30 days over Europe. Mission length: 15 years.	19 (contract signed April 2008)	2012
	SENTINEL-3	Dedicated to oceanography and global monitoring of land vegetation. Sentinel-3 will determine parameters such as sea surface topography, sea/land surface temperature, ocean colour and land colour. For this purpose, it carries an advanced radar altimeter and a multi-channel optical imaging instrument.	305 (contract signed April 2008)	2012
	SENTINEL-4 and 5	Due to be dedicated to meteorology and atmosphere studies.	TBD	TBD

1. Core Explorer missions address specific areas of scientific interest (cost cap: EUR 300-400 million).
2. Opportunity Explorer missions: faster, lower-cost missions to address immediate environmental concerns (cost cap: EUR 150 million). As of June 2008, there were six additional proposals for core missions undergoing assessment study.
Source: Based on ESA website and OED/IFP research.

requirements for the next decade or so – averaged out over the full time frame – would not seem to be out of line with the investments identified for the next decade in the previous section and in recent market studies (Adam, 2008).[7]

It cannot be taken for granted that such levels of investment in space systems will automatically be forthcoming. They are subject to difficult, often contentious political, technical and economic decision-making processes. What will be required to ensure that adequate levels of investment in space systems are ultimately realised? The answer, first and foremost, is an improved and extended tool box to help policy makers arrive at investment decisions based on good data and analysis. Such a tool box is offered in the next chapter.

Notes

1. Not all the geostationary satellites are operating in full mode; some are in "standby" mode, in orbit to take over if another satellite malfunctions.

2. Useful status information, updated on a regular basis, can be found on the joint CGMS (Coordination Group for Meteorological Satellites) web pages: *http://cgms.wmo.int*.

3. The 1975 Convention on Registration of Objects Launched into Outer Space obligates countries to keep a registry of what they launch and file it with the United Nations; however, few nations contribute to the registry, and often with delays. The United States NORAD (North American Aerospace Defense Command) also has a registry of space objects in orbit (including space debris), which is used in some commercial products.

4. ASPRS lists 102 earth observation satellites in orbit as of February 2006 (*www.asprs.org/news/satellites/*). The think tank Union of Concerned Scientists lists 107 remote sensing satellites in its database (*www.ucsusa.org/global_security/space_weapons/whats-in-space.html*). NOAA provides a list of optical satellites (36 as of end-2006) and radar satellites (4 as of end 2006) (*www.licensing.noaa.gov/whatsnew2.html*). ESA's eoPortal Directory provides a comprehensive collection of earth observation-related resources; over 200 satellites are listed, not all operational (*http://directory.eoportal.org/*).

5. All Sentinel missions should have an operational lifetime of 7 years, with a possible extension of 5 more years.

6. MetOp-A was launched in October 2006 as Europe's first polar-orbiting satellite dedicated to operational meteorology. Another ESA-EUMETSAT programme is under way for future satellites in geostationary orbit (*i.e.* Meteosat Third Generation or MTG). MTG will take the relay in 2015 from Meteosat 11, the last of a series of four satellites of the MSG (Meteosat Second Generation).

7. According to a recent Euroconsult report, the world manufacturing market for earth observation satellites – not including meteorological satellites that are integrated in the OECD analysis – could grow 14%, from USD 14.8 billion over the 1997-2006 period to USD 16.9 billion over 2007-16 (Adam, 2008).

Bibliography

Adam, Keith (2008), "Satellite-based Earth Observation Entering Expansion Phase", *EoMag*, European Association of Remote Sensing Companies (EARSC), March.

Alverson, Keith (2007), "Why the World Needs a Global Ocean Observing System", *Marine Scientist*, No. 21, pp. 25-28.

Cauzac, J.P. and J.Y. LeBras (2008), "What Space Technologies Have Really Changed In Maritime Security", *Proceedings "Space Applications 2008"*, Toulouse Space Show, 23-25 April.

CEOS (Committee on Earth Observation Satellites) (2007), *Satellite Observation of the Climate System: CEOS Response to the Global Climate Observing System (GCOS) Implementation Plan (IP) 2006*, September.

Chengyi, Z. (2008), "China's Earth Observation Activities", Second GEOSS Asia-Pacific Symposium, Tokyo, Japan.

Coletta, Alessandro *et al.* (2008), "COSMO-SkyMed Mission status", Proceedings of the SeaSAR Workshop 2008, European Space Agency, Frascati, Italy, 21-25 January.

Eyre, John (2008), *Vision for the GOS: Review Inputs to Preparation of Revised Vision of the GOS in 2025*, Report Submitted by Dr John Eyre, ET-EGOS Chairperson, UK Met Office, CBS/OPAG-IOS/ET-EGOS-4/Doc. 10.1 (11.VI.2008), Commission for Basic Systems, Open Programmme Area Group on Integrated Observing Systems, Expert Team on Evolution of the Global Observing System, Fourth Session, World Meteorological Organisation, Geneva, Switzerland, 7–11 July.

EC / JRC (European Commission/Joint Research Centre) (2008), "Integrated Maritime Policy for the EU", Working Document III on maritime surveillance systems, European Commission/Joint Research Centre, Ispra, Italy, 8 January.

GCOS (Global Climate Observing System) (2005), *Analysis of Data Exchange Problems in Global Atmospheric and Hydrological Networks*, GCOS-96, February.

Høye Gudrun K., *et al.* (2008), "Space-Based AIS For Global Maritime Traffic Monitoring", *Acta Astronautica*, 62, pp. 240 – 245, January-February.

ICAO (International Civil Aviation Organization) (2005), *Global Navigation Satellite System (GNSS) Manual*, Document 9849, 1st Edition.

ICG (International Committee on Global Navigation Satellite Systems) (2008), Proceedings of the Second Meeting of the GNSS Providers Forum, Vienna, Austria, 18 February. Website: *www.unoosa.org/oosa/en/SAP/gnss/icg.html*.

Lafeuille, Jérôme (2008), "Geostationary Satellites in a World Meteorological Organization Perspective", Presentation made at the 5th GOES Users' Conference, New Orleans, 23-24 January.

Milsom, Edward (2008), "The Sky Is No Limit: Norwegian Satellites Show the Way", Nortrade, 27 May.

NRC (National Research Council) (2007), *Earth Science and Applications from Space: National Imperatives for the Next Decade and Beyond*, Space Studies Board (SSB), Washington DC, January.

OECD (2005), *Space 2030: Tackling Society's Challenges*, OECD, Paris.

Russian Space Agency (2008), *GLONASS Constellation Status*, Website: *www.glonass-ianc.rsa.ru*, accessed 10 May.

USGS (US Geological Survey) (2008), "Imagery for Everyone", *USGS Press Release*, 19 April.

US National Academies of Sciences (2007), *Earth Science and Applications from Space: National Imperatives for the Next Decade and Beyond*, Washington DC.

WMO (World Meteorological Organization) (2002), *Observational Data Requirements and Redesign of the Global Observing Systems*, Commission for Basic Systems Extraordinary Session, Cairns, 4-12 December.

WMO (2007), *Final Report: WMO Workshop on the Redesign and Optimization of the Space-based GOS*, WMO Headquarters, Geneva, 21-22, June.

WMO (2008), World Weather Watch Programme, Website: *www.wmo.ch/pages/prog/www/index_en.html*, accessed 2 May.

ISBN 978-92-64-05413-4
Space Technologies and Climate Change
Implications for Water Management, Marine Resources
and Maritime Transport
© OECD 2008

Chapter 6

Tool Box for Policy Makers: Costs, Benefits and Investment Decisions

Investment in space systems must be sustained in order to meet the challenges of climate change, natural resources management and activities that affect the environment, such as maritime transport. However, it cannot be taken for granted that the funding will be forthcoming, not least because cost-benefit evaluations of future investments are far from straightforward. This chapter suggests how improvements might be made to the policy makers' tool box for assessing and deciding on space-related investments. The first section discusses some key notions useful for defining space programmes (e.g. R&D, infrastructure, notion of costs); a second reviews methodologies used to evaluate benefits of space programmes in more detail, using specific case studies as illustrations; a third section develops an innovative infrastructure approach; and a final section provides prospective views on the use of risk management approaches for decision making in the field of space investments.

Table 6.1. **Main evaluation methods for analysing large programmes**

Selected methods	Description
1. R&D Programmes' Impact Analysis	
Scientific returns	Quantifiable measure of publications
Economic returns	Quantifiable parameters to try and link R&D intensity and economic activity (e.g. technology transfers)
2. "Classic" return on investment techniques:	
Key performance indicators	Quantifiable performance measures
Cost-benefit analysis (CBA)	Measures tangible and intangible benefits and assesses these against costs
Break-even analysis	The amount of time necessary for benefits to equal costs
Transaction costs	Segmentation methods to calculate use and benefits to different user groups
Cost-effectiveness	Marginal costs for achieving specific goals
Net present value	The difference between the present value of cash inflows and outflows at a given discount rate
Initial rate of return	The discount rate that makes net present value of all cash flows equal to zero
Value assessment	A method that captures and measures factors unaccounted for in traditional return on investment (ROI) calculations
Portfolio analysis	A method that quantifies aggregate risks relative to expected returns for a portfolio of initiatives
Real options analysis	Analysis of capital investments in terms of the options they contain, with uncertainty accounted for by risk-adjusting probabilities ("equivalent martingale approach")
3. Infrastructure approach	Benchmarking investments in space systems against terrestrial infrastructure investments
4. Risk management approach	Addresses investments in satellite systems from the point of view of monitoring and mitigating major risks and reducing uncertainties

Source: Adapted from OECD, 2006b.

Definitions and key notions for evaluating space programmes

In most OECD member countries, governments increasingly require that public agencies assess the benefits and costs of their operations while exploring the possible monetisation of these benefits. For the past 20 years the space-related agencies and industry, particularly in Europe, North America and India, have examined ways of estimating benefits from space programmes (Sankar, 2007). There is a general sense that over time, the cost of launching and operating a satellite is offset by the many benefits it provides; nonetheless, measuring those benefits is a challenge. In the case of specific applications that use space assets in large or small measure, such as water management or maritime zone control, there is added complexity intervening in the final decisions (*e.g.* about the value of satellite data, gauges on the ground, specific actions led by actors on the ground such as civil protection in case of floods). This section reviews some definitions and key notions for evaluating space programmes.

Definitions

Space programmes provide an interesting paradox: they are often considered and funded as research and development (R&D) programmes, but act in many cases as key infrastructures delivering unique public and private services. This subsection reviews a number of terms.

Space programmes – The first full-scale space programmes date from the late 1940s – early 1950s. From the start, they consisted of R&D projects to develop technologies and know-how to send objects into space and utilise this new dimension for science and security purposes. Today, institutional space programmes worldwide still cover a wide range of technologies (*i.e.* launchers, satellites, space stations, ground segment) and disciplines (*e.g* telecommunications, earth observation, navigation, astronomy), sometimes with "accompanying" programmes to involve new users (*e.g.* commercialisation of technologies outside the space sector). Space programmes are usually undertaken nationally via dedicated agencies, but also often within a bilateral or multilateral international co-operation framework, particularly in the European context. Since the 1980s, a number of private actors have conducted their own space programmes directly, for profit (*e.g.* telecommunications satellite operators, commercial launch providers), but always within a regulatory framework put in place by governments (OECD, 2005).

Space applications – "Applications" are the resulting outcomes of many space programmes. Sometimes they are actively sought, to develop specific space products and services (*e.g.* satellite television); on occasion such results are accidental. The data derived and/or signal issued from a large number of programmes initiated for purely scientific purposes can be deemed relevant by large communities of users. Today the value chains for space applications vary, depending on the commercial or scientific benefits of the data or signal provided. That is where the destination between pure R&D programmes (set up for a limited period) and applications (often to be set up on an enduring operational basis) becomes blurry at times. Even in the case of military space programmes, it has been historically difficult to shift programmes from the science and technology environment to the operational environment (GAO, 2008a).

Space infrastructure – The term "space infrastructure" encompasses all systems, whether public or private, that can be used to deliver space-based services. These include both the space and underlying necessary ground segments. As identified in OECD (2005), there are two complementary and interlinked space-based infrastructures. The first one focuses on the "front office", i.e. the one "user-oriented" and designed to provide information-related services including communications, navigation signals and earth observation data to governments and society at large. The second concerns the essential enabling "back office", i.e. the space transport, satellite manufacturing and servicing infrastructure. This notion of space infrastructure will be explored further in the chapter.

Assessing the costs

In order to assess the net benefits of a programme, it is essential to have an idea of the costs of developing it. Producing a comprehensive cost analysis of any space system can be daunting. Historically, major space projects have tended to be markedly complex and lengthy and therefore costly, even in the case of commercial satellites. Space systems remain high-level technological products, the results of long-term and constant research and development. Even when a satellite platform is "standardised", furnishing a prototype that can be reused for other missions, the addition of new instruments or sensors adds development costs and complexity. Recent advances in small satellites have alleviated some costs, but even then the overall budget for a programme may not be easily envisaged (OECD, 2004). This is especially true for expected operations costs, which may be much larger than previously anticipated. Such is the case with an unexpectedly increased satellite lifetime (e.g. the European ERS-2 satellite, designed for three years in orbit but still operational today after more than 12).

Three main categories of costs are usually estimated in any R&D project:

- The direct costs of a new system or integration initiative (e.g. financial costs), which are usually the easiest to identify and analyse.

- The opportunity costs, which are the losses or costs to the organisation that result from developing a new system rather than using the resources on alternative projects (the engineer who spends several hours learning a new computer system to prepare for a new project instead of working on another short term project, for example, has incurred a limited opportunity cost).

- The indirect costs, such as those for infrastructure maintenance, depreciation or the overall administration expenses, are usually based on uncertain assumptions and limited knowledge of actual impact of shared resources.

Three main techniques are used to estimate these costs in most high-technology sectors, including in the space sector during a project's life cycle (i.e., research and development, production, operations and support costs). These methodologies, along with their advantages and disadvantages, are summarised in Table 6.2. Parametric modelling, used for brand new systems, is particularly challenging, as there are no traceable historical data (Glad, 2005).

Table 6.2. **Cost estimation techniques for space projects**

Method	Characteristics	Advantages	Disadvantages
Cost by analogy	Derived from costs taken from a technically similar programme	Costs traceable to historical data	Not sensitive to programmatic details
Parametric modelling	Derived from relationship of costs to physical data (*e.g.* weight, power consumption)	Provides a rough order of magnitude estimate (*i.e.* specific cost drivers) if the system characteristics are already defined	Not traceable to historical data, and some cost drivers can be underestimated because of innovative combinations of subsystems
Engineering modelling	Detailed information and cost element estimates prepared at the lower practical level of task design and definition (*e.g.* work breakdown structure, or WBS)	Can use the work breakdown structure elements' historical cost data	Requires a very detailed design description from the start; time consuming and not possible for very innovative projects

Source: Adapted from Cohendet, 1999.

Space systems often differ greatly, and despite the hundreds of satellites sent into orbit since the 1950s they have so far generally been produced in rather small quantities each time. Many environmental satellites can even be qualified as prototypes. The technological risks also represent an inherent difficulty, which may cause costly delays (Box 6.1). Cost estimates for space systems and their derived applications are thus a domain in which it is difficult to generalise.

Tracing benefits to satellites

As seen in Chapter 4, satellites have some unique capabilities. But it can be difficult to trace the benefits back to space-based systems that provide specific links, signals or data. It is even more complex to then try and pinpoint the benefits of a specific instrument or sensor on one satellite.

As an example, the satellite MetOp 1 is Europe's first polar-orbiting satellite dedicated to operational meteorology. MetOp carries onboard 11 instruments, composed of a set of "heritage" instruments provided by the United States and a new generation of European instruments that offer improved remote sensing capabilities for many different disciplines.The new instruments already increase the accuracy of temperature humidity measurements, readings of wind speed and direction, and atmospheric ozone profiles.[1] However, improved weather forecasting products and environmental monitoring systems could be affected as much by improvements in data thanks to MetOp as by better modelling techniques, which are in full development.

By the same token, a wide range of geophysical parameters can be derived from a single satellite instrument. The Special Sensor Microwave Imager (SSM/I) was mentioned several times in Chapter 4. This sensor is carried onboard several American meteorological satellites (*i.e.* F13, F14 and F15), allowing 24-

---

Box 6.1. **Challenges in evaluating space programme costs**

The case of the US Department of Defense's space acquisition programme

The US Department of Defense (DOD) invests heavily in space assets to provide the armed forces with intelligence, navigation and other information critical to conducting military operations. In fiscal year 2008 alone, the DOD expects to spend over USD 22 billion on space systems. The majority of major acquisition programmes in the DOD'S space portfolio have, however, experienced problems during the past two decades that have driven up costs and schedules and increased technical risks. At times, cost growth has come close to or exceeded 100%, causing the DOD to nearly double its investment without realising a better return. Along with the increases, many programmes are experiencing significant schedule delays, as much as seven years. A number of reasons for these problems have been identified by the Government Accountability Office (GAO). They include optimistic cost and schedule estimating; the tendency to start programmes with too many unknowns about technology; inadequate contracting strategies; contract and programme management weaknesses; the loss of technical expertise; capability gaps in the industrial base; tensions between laboratories that develop technologies for the future and acquisition programmes; the different needs of users of space systems; and diffuse leadership. The DOD is taking a number of actions to address the problems reported by the GAO.

Source: US Government Accountability Office (GAO), 2008b.

---

hour coverage. Its data are used to study a myriad of elements: ocean surface wind speed, an area covered by ice, the age of ice, ice edge, precipitation over land, cloud liquid water, integrated water vapour, precipitation over water, soil moisture, land surface temperature, snow cover and sea surface temperature. If one includes its predecessor (the Scanning Multichannel Microwave Radiometer SMMR carried on board several meteorological satellites), SSM/I data have in fact been available from late-1978 to the present. This is a case where is one instrument has been providing valuable data – for decades – to a variety of scientific and operational users. Other sensors with the same versatility, and which can be used to observe other parameters than those for which they were originally designed, include for instance the GOME-2 spectrometer. Developed to monitor ozone, the sensor can also be used for water vapour detection. As another example, soil moisture, ice and snow can be monitored by ASCAT-MetOp, whose original purpose was wind measurements.

One method used to discriminate among various sensors' contributions on satellites was attempted in the ROSE GSE study (Whitelaw *et al.*, 2004). The idea was to determine which data were the most likely used by an application, and estimate the number of pictures that may needed by a specific user to conduct his activity. The methodology used was to take the marginal cost of the proposed/future missions, then estimate a proportional cost of the missions based on the levels of use of a specific instrument in proportion to overall mission data use. This determined an appropriate fraction of the total mission cost. As mentioned by the authors (Whitelaw *et al.*, 2004), the method is valuable as it is a move towards addressing the real cost of satellite-based information. The estimate nevertheless remains very approximate given the difficulties of estimating mission costs (including satellite construction, launch and operations); levels of use in proportion to overall total usage; shared use of given acquisitions; and actual mission lifetimes. The results of this type of study could usefully be transferred to other application studies using benefit transfer methods and clear caveats. The lingering problem is that the overall study ignores by definition the overall costs of the earth observation systems, as the prices of satellite imagery – even if still high in many cases – do not reflect the total costs of setting up and operating the infrastructure.

Table 6.3. **Cost base for the ROSE study (2004)**

Mission/instrument	Mission Cost (million euros)	Duration years	Annual cost	Instr. % of mission	Annual instrument cost	Duty cycle	Ops factor	Images per year	Cost per scene (euros)
ENVISAT ASAR	2 300	5	460	50	230.00	10	75.0	157	680
ENVISAT MERIS	2 300	5	460	20	92.00	100	25.0	525	600
ENVISAT ATSR	2 300	5	460	5	23.00	100	50.0	1 576 800	15
ERS 1/2	500	5	100	50	50.00	10	75.0	157 680	317
NASA MODIS (Terra and Aqua sats)	1 130	5	226	20	45.20	100	25.0	525 600	86
NOAA AVHRR	202	5	40	60	24.24	100	90.0	1 892 160	13
Radarsat-1 FB,IM,WS	426	5	85	100	85.20	10	75.0	157 680	540
SPOT HRS, HRG	671	5	134	80	107.36	100	15.0	315 360	340
Orbview 2 SeaWiFS	43.5	8	5	100	5.44	100	25.0	525 600	10
Radarsat-2	380.25	5	76	100	76.05	10	75.0	157 680	482

Source: Whitelaw, *et al.*, 2004.

Timeliness of studies

Technological advances can affect both costs and benefits. Cost-benefit studies may often become quickly out of date, as technology evolves and the cost efficiency and capability of systems improve over time. For example, there have been rapid advances in satellite communications and earth observation technologies in recent years. This means that more can be done (*i.e.* higher benefits) for less (*i.e.* lower costs) with the latest generation of earth observation satellites, and future generations could be even more effective. Another aspect deals with the maturity of services at least linked partially to space-based data. As seen in Chapter 4, most of the sensors that were identified for climate and water management missions are for research purposes. As such, they aim to demonstrate capabilities and may have a rather short time scale. Prospective views are therefore often needed, and render the cost-benefit exercises even more difficult.

Diminution of recognised benefits due to external factors

Concerning the use of new technologies, the general purposes for which data or signals are used can directly affect the expected benefits from a system. This can of course relate directly to the role of space systems and climate change. If some aspects of climate management receive undue attention or inaccurate appreciation, that might in turn diminish recognition of the anticipated benefits derived from space systems. Conversely, justification for new technologies and claims as to their efficacy may lack critical scrutiny. That either situation is possible indicates a need for the appraisal process to involve a full, close, and systematic examination of claims concerning the benefits of a technology or product. And that includes identifying and assessing the conditions under which the claimed benefits might or might not materialise.

Measuring the impacts of R&D programmes

The current assessment efforts of public science and research programmes in general still fail to capture the full range of impacts derived from R&D. As identified by OECD (2007e), the methodologies used are still evolving. Moreover, the choice of a specific analytical technique for impact assessment is not random but context-specific. The timing and objective of the assessment, as well as the nature and scope of the public R&D funded, are factors that must be borne in mind when selecting an analytical technique from an existing toolbox. Assessment of space programmes is also affected by those challenges, as will be shown in this section. A first subsection looks at scientific return measurement issues, and a second reviews macro and micro approaches used when assessing economic returns.

Scientific returns from R&D programmes

Science has historically been a major objective of space programmes. In terms of scientific returns, one useful quantitative measure concerns the number of refereed publications based on space missions. However, the true impact of a mission could go significantly beyond the publications that specifically mention it, as new research and operations techniques are often developed in the framework of a given mission.

In the case of the Tropical Rainfall Measuring Mission (TRMM) for example – the mission mentioned earlier between NASA and the Japan Aerospace Exploration Agency (JAXA) – the launch triggered a flood of research that greatly broadened understanding of tropical weather systems and their forecasting, as well as improved quantification of the hydrological cycle and the climate system (NRC, 2006).

The studies captured in Figure 6.1 span a broad spectrum of topics. They include contributions to increasing the basic scientific knowledge needed for future applications (*e.g.* descriptive and diagnostic studies) as well as operational applications (*e.g.* monitoring weather features, notably tropical cyclone activity; climate monitoring; numerical weather prediction and climate model development; and model assimilation of TRMM data in forecast operations). Papers addressing operational aspects (lower curve) lag behind scientific papers (higher curve), but that will change as operational use of the TRMM data substantially increases (*i.e.* as quality control issues are resolved and data are progressively integrated in applications).

Figure 6.1. **Scientific returns measurement of refereed publications directly related to TRMM**

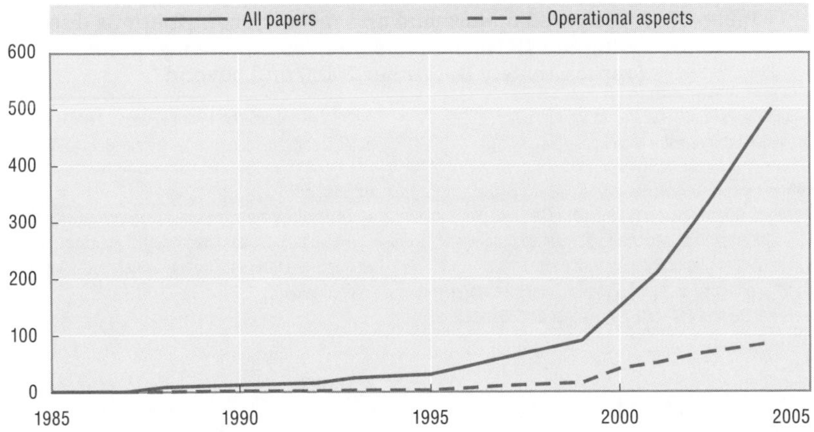

Note: Data for the figure were obtained by the referenced authors through searching the *Institute for Scientific Information's Science Citation Index* for papers that mention TRMM either in the title, abstract or keywords. Papers dealing with operational aspects are based on terms such as "real-time,""operational" and "assimilation."

Source: Matthias Steiner, Princeton University, cited in NRC, 2006.

It is challenging to anticipate scientific benefits for most first-generation satellite sensors due to a lack of prior experience with similar data (NASA, 2002). But often the planning and forecasting of a satellite mission lifetime hinge on those potential scientific returns. Taking into account this uncertainty, space agencies' resource planning for most earth observation missions has focused on a relatively short-term, high-payoff approach based on expected fundamental scientific and engineering returns. This has caused considerable dissatisfaction in the climate science community, where long-term observations are essential to fundamental research.

As an example, the success of the Tropical Rainfall Measuring Mission and its continuing "good health" led to several extensions of the mission.

Conclusions about the benefits of extending TRMM to and beyond the "fuel point" (the maximum time when the satellite's re-entry in the atmosphere can still be safely controlled) are compiled in Table 6.4. Although the additional cost of extending TRMM from December 2004 to November 2005 was estimated at approximately USD 4 million,[2] the many derived intangible benefits of pursuing operations seemed prior to the extensionto outweigh the costs (NRC, 2006). Since 1998, TRMM has in particular provided near-real-time information for operational purposes (behaviour of tropical cyclones, rainfall), and no other satellite can replace it for the moment. The TRMM mission was extended again in 2005; it is expected to last until at least 2010 (as of June 2008), when the first of a series of planned follow-on Global Precipitation Measurement Mission satellites is due to be launched.

Table 6.4. **Anticipated operational and research contributions due to extending the TRMM satellite missions to the fuel point (approximately December 2005) and beyond**

Anticipated contributions of TRMM up to the fuel point (when controlled TRMM re-entry is still possible)	Additional anticipated contributions of TRMM beyond the fuel point (*i.e.* in addition to what is gained up to the fuel point)
OPERATIONS	**OPERATIONS**
• Another year of TRMM Microwave Imager (TMI) and precipitation radar (PR) data enhancing near-real-time rainfall products** • Another year of lightning data for air traffic advisories* • Realising PR's potential as a global rainfall reference standard* • Another year of PR and TMI data for weather and climate prediction models**	• Technology demonstration of the endurance of the first precipitation radar forecasting** in space • Improved forecasts from the operational numerical weather prediction**
RESEARCH	**RESEARCH**
• Overlap with CloudSat radar operations and the A-Train satellite experiment ** • Overlap with the Coriolis WindSat sensor mission* • Unique opportunities to enhance field experiments (TCSP, TEXMEX-II)** • Unique opportunities to enhance international research programmes (GEWEX, THORPEX, Hurricane Field Program)** • TRMM's precipitation radar provides calibration reference for the current global precipitation measurement mission-like Oscillation cycle* constellation of microwave satellite sensors** • TRMM is a catalyst for tropical cyclone research (*e.g.* research on convective bursts, tropical cyclone eye wall replacement cycles, improved forecasting of inland flooding during hurricanes)** • Longer TRMM record needed for tropical cyclone forecasting* • Longer TRMM record needed for climate research* • Foster improvement in moist physics parameterisation for climate models, numerical weather prediction, and related assimilation systems by evaluating models of clouds and precipitation physics*	• Unique opportunities to enhance field series** • Developing the next generation hurricane forecast model** • Better characterisation of interannual variability and the El Niño-Southern Oscillation cycle* • Seamless transition into global precipitation measurement (GPM) operations* • Realisation of a GPM-like prototype * Avoiding researchers being ill-prepared for future global precipitation measurement operations**

Note: A single asterisk differentiates applications that use TRMM data as the only or primary component of a research or operational activity from those that use TRMM data as a complementary component (marked with a double asterisk). There is a grey area between these two categories, but the distinction serves as a first-order attempt to differentiate between essentially stand-alone contributions and complementary but still unique contributions of TRMM.
Source: NRC, 2006.

Economic returns of R&D programmes

In addition to evaluating scientific returns, different methods derived from economic theory using macro and micro analysis techniques have been used over the years to assess the economic effects of space technologies. These are detailed below with references to specific studies.

i) The macroeconomic approach

Economic growth theory has long postulated that improvements in technology are the source of long-run development (Solow, 1956; Romer, 1990) and that differences in technology are the main determinant of differences in income per capita across countries. The macroeconomic approach is often used in the case of large R&D programmes or infrastructure to provide cost-benefit information, via economic input-output analyses. The main objective is to measure the growth of productivity in a region or country generated by the investment.

Input-output analysis specifically shows how industries are linked together through supplying inputs for the output of an economy. Factors that can be used to construct indicators of productivity are for example employment, expenditures, income, production of goods and services and competitiveness. Such factors are of interest at both the national and regional levels. Results of these analyses are derived from macroeconomic data such as changes in GDP, which can then be compared to changes in capital. The challenge when interpreting the material is to find the causal linkages between the programme/ infrastructure investments and the rise in productivity. However, the findings of these studies are sometimes contentious, and highly dependent on the choice and evaluation of appropriate variables over long periods, as well as the calculations used to assess their cause and effect mechanisms.

As an example, the Federal Aviation Administration's Office of Commercial Space Transportation (FAA/AST) published a report in 2006 on the impacts of commercial space transportation and related industries in other economic sectors, specifically in terms of revenues and jobs that are generated (FAA, 2006). The economic impact analysis used an input/output method and the Regional Input-Output Modelling System (RIMS II) developed by the Department of Commerce, Bureau of Economic Analysis. The space sector, as defined in the study by the FAA, was found to be responsible – via direct, indirect and induced impacts – for USD 98 billion in economic activity in 2004 and 551 350 derived jobs throughout the United States. All major US industry sectors were affected positively to some extent (e.g. the information services sector, manufacturing, finance and insurance, healthcare and social assistance). As a comparison, using the same methodology the economic impact of the civil aviation industry was found to be over 10 times that of commercial space transportation and enabled industries. Methodology-wise, input-output analyses are valuable methods to measure economic impacts. On the other hand, one inevitable drawback of this type of analysis stems from the lack of precise space statistics, since the statistical codes used for the study by definition cover more than just space activities (OECD, 2007f).

Other studies have been conducted using input-output techniques to study the macroeconomic impacts of space programmes at local or regional levels. Whenever many employees from a single organisation are working in one area, it is assumed that economic spillovers can be felt in a given region (the same concept applies to the economic effects of large military bases). As an example, with more than 1 600 NASA scientists and engineers, the John C. Stennis Space Center strongly influences the surrounding communities. In 2005, the NASA centre's direct global economic impact was estimated at a total of USD 691 million, with a USD 503 million impact on Mississippi and Louisiana communities within a 50-mile radius. Impact studies have also been conducted for French Guiana, host of "Centre Spatial Guyanais", the European spaceport. Onsite space activities represent 20% of the French department's GDP in 2005, with 1 350 persons employed in the sector and 5 800 derived jobs in other sectors (one direct job being responsible for 4.4 induced jobs). In addition, actors involved in the space sector are responsible for 40% of local taxes and 60% of French Guiana imports (CNES and INSEE, 2005).

The microeconomic approach

Microeconomic analysis studies the behaviour of individual organisations, firms and customers and their interactions, usually determined by market demand and supply. The use of supply and demand curves is however not always directly applicable to space systems and their derived applications because of immature products (new technologies) and non-quantifiable demand. A real technical limitation of microeconomic analysis is the daunting task of assessing accurately all the markets liable to be affected by a specific space technology, and not just when it is innovative (Eurospace, 1994). Different microeconomic approaches are presented below.

Numerous studies of "spin-offs" have been conducted in the United States since the 1960s (such as outputs from NASA's Apollo programme), notably of the transfers from space-related hardware and know-how to other sectors (e.g. medical imagery). The value of spin-offs is however not easily quantifiable, and at times oversold concepts have been detrimental (e.g. Teflon as space technology).

In Europe, the BETA (*Bureau d'Économie Théorique et Appliquée*) of the University Louis Pasteur of Strasbourg has developed over the past 20 years a methodology extensively applied by them to assess the indirect economic effects of ESA contracts in European member countries (BETA, 1989, 1997; Bach, 2002).

The method focuses on the indirect effects generated by an R&D project such as a space programme or a space contract with a commercial firm. Those effects generally occur in four areas: technology (*e.g.* sales of the same product to other customers, or improvement in the current product line based on the space technology developed); marketing (*e.g.* reputation and image enhancement resulting from the execution of the contract for ESA); organisation and method (*e.g.* better performance thanks to standards and management techniques used during the ESA contract); and critical mass (*e.g.* preserving or increasing the number of employees in the space sector). The effects are identified and measured through extensive interviews with personnel in each firm that received a contract over the period studied. Using the same methodology over different periods, BETA measured the effects of ESA contracts in Canada and European countries, with generally positive findings for the space sector. But as mentioned in Amesse and Cohendet (2001), "most of the effects are concentrated within the space sector and are increasing over time as specialisation of the industry increases. There is also a tendency for effects to be concentrated within contracting firms, leaving less and less room for subcontractors." Investment in the space sector has traditionally remained in the space sector. Even for product development, it is shown that technological effects "are mainly short term (improved products and resale of space products to other foreign customers); while the development of new products is a less important factor and decreases over time." This trend could be expected due to the very specific nature of space-related manufacturing, but the increasing number of downstream applications in numerous disciplines and uptake of space-based data could signal change.

A combination of macro and micro approaches could provide better estimates, although it will still fail to address potentially larger non economic impacts (Sankar, 2007). The Belgium federal government commissioned a study to researchers at the Université Libre de Bruxelles (ULB) to evaluate the social and economic contribution of the federal scientific research institutes linked to the space programme (Capron and Baudewyns, 2007). Despite a relatively small space budget compared with the major space powers, Belgium ranked eighth in 2005 in terms of national public space budgets as a percentage of GDP (OECD, 2007f). The Belgian space sector employs some 1 600 scientists, engineers and technicians (Beka, 2007). In order to calculate those contributions, the ULB is planning to use both input-output tables and a large survey to identify and quantify the economic impacts derived from three institutes.

Judging from the examples presented so far, large-scale R&D programmes in the space sector have been the focus of many socio-economic studies over the years. However, all such studies face inherent limitations, very similar to those in other types of public R&D impacts analysis (Box 6.2). When assessing the results of these studies there is often a reluctance to link socio-economic outcome measures too directly to research programmes, as there are many

Box 6.2. **Challenges encountered when analysing the impacts of public R&D**

- **Causality.** There is typically no direct link between a research investment and an impact. Research inputs generate particular outputs that will then have an impact on society. As it is indirect, this relationship is difficult to identify and measure. It is also almost impossible to isolate the influence of one specific factor (research output) on one impact, because the latter is in general affected by several factors that are difficult to control for.

- **Sector specificities.** Creation and channelling of output to the end-user will differ depending on the research field and industry. This renders ineffective the use of one single framework for assessment.

- **Multiple benefits.** A basic research impact may have several dimensions, not all of which are easily identified.

- **Identification of users.** Identification of all end-users who benefit from the research outputs can be difficult and/or costly, especially in the case of basic research.

- **Complex transfer mechanisms.** It is difficult to identify and describe all the potential mechanisms for transferring research results to society. Some studies have identified mechanisms of transfer between businesses or between universities and businesses. These models are mainly empirical and often reveal little of the full impact on society of such transfers.

- **Lack of appropriate indicators.** Since appropriate benefit categories, relevant transfer mechanisms and end-users are often lacking, it is also difficult to define and measure appropriate impact indicators related to specific research outputs.

- **International spillovers.** The existence of knowledge spillovers has been well documented and demonstrated. As a result, specific impacts could be partially the result of internationally performed research instead of national research investments.

- **Time lags.** Different research investments vary in the time it takes them to have an impact on society. Any measurement may thus prove premature, especially in the case of basic research.

- **Interdisciplinary output.** Research outputs, *e.g.* improved skills, may have different impacts, and it may be difficult to identify them all in order to evaluate the contribution of the specific output, let alone that of the research investment.

- **Valuation.** In many cases it is difficult to come up with a monetary value of the impacts so as to make them comparable. Even if identifiable, noneconomic impacts may be difficult to value. There have been some attempts to translate some of these impacts (such as the economic savings associated with a healthy population or the calculation of opinion values) into economic terms, but these have typically remained partial and open to subjectivity.

Source: OECD, 2007e.

intervening steps that may distort the causal link. The next sections present other return on investment approaches (direct and indirect market valuation techniques), notably using cost-benefit analysis.

"Classic" return on investment techniques (direct and indirect market valuation)

The most common economic measurement of any technology's value is the calculation of return on investment (ROI), a ratio of costs to benefits. To calculate the ROI of a space system, it is necessary to total the costs of deploying the system (*e.g.* hardware, software, maintenance, training and so forth) – and divide the total by the potential benefits (such as improved productivity, decreased cost of operations, increased revenue and better customer satisfaction rates in the case of commercial systems). However, as mentioned in previous sections, space systems are by nature very specific, due to their complexity, risks and lengthy research and development. There is extensive academic literature on cost-benefit analysis (CBA), also sometimes called "project appraisal" or "policy evaluation", for many current societal interests. Methodologies for benefit assessment continue to evolve.

The challenge of placing monetary value on technologies and the services they deliver is not just a space problem. Whether cost-justifying new purchases, assessing the value of existing assets or making a business case for future architectural directions, valuing a technical system is a complex and often subjective exercise. Monetary or financial valuation methods fall into three basic types, each with its own repertoire of associated measurement issues (Table 6.5); not all of them can be used to estimate satisfactorily the benefits derived from space systems. They are:

- Direct market valuation.
- Indirect market valuation.
- Survey-based valuation (contingent and group valuation).

Direct market valuation techniques

Direct market valuation techniques are the obvious methods for assessing established commercial space applications, since the economic importance of a service can be directly measured in monetary units. But value can also be extrapolated by the service contribution to employment and productivity, *e.g.* in terms of the number of people whose jobs are related to the use or conservation of the service.

Market price – This classic method uses exchange value, based on the marginal productivity cost that products and services have in trade. It is mainly applicable to well-identified commercial space-based products and services (*e.g.* phone conversation using a satellite link).

Table 6.5. **Monetary valuation methods, constraints and examples**

	Method	Description	Constraints	Examples
1. Direct market valuation	Market price	Measures the exchange value (based on marginal productivity cost) that products and services have in trade	Market imperfections and policy failures distort market prices	Mainly applicable to identified commercial products and services (*e.g.* phone conversation using a satellite link), "goods", and some cultural (*e.g.* recreation) and regulating services
	Factor income or production factor method	Measures the effect of services on loss (or gains) in earnings and/or productivity	Care needs to be taken not to double-count values	Number of people whose jobs are related to the use of specific space service (*e.g.* telecommuting via satellite communications); for ecosystem analysis: natural water quality improvements that increase commercial fisheries catch
	Public pricing	Measures public investments, *e.g.* land purchase, or monetary incentives (taxes/subsidies) for ecosystem service use or conservation	Property rights are sometimes difficult to establish; care must be taken to avoid perverse incentives	Investments in watershed protection to provide drinking water, or conservation measures using satellite monitoring systems
2. Indirect market valuation	Avoided (damage) cost method	Services that allow society to avoid costs that would have been incurred in the absence of those services	It is assumed that the costs of avoided damage or substitutes match the original benefit. However, this match may not be accurate, which can lead to underestimates as well as overestimates	The value of a satellite-based flood control service can be derived from the estimated damage if flooding were to occur.
	Replacement cost and substitution cost	Some satellite services could be replaced with other systems		The value of a satellite system can be estimated from the costs of obtaining information from another source (substitute costs)
	Mitigation or restoration cost	Cost of moderating effects of lost functions (or of their restoration)		Cost of preventive expenditures in absence of communication services (*e.g.* lives potentially lost)
	Travel cost method	Use of ecosystem, water-related services may require travel and the associated costs can be seen as a reflection of the implied value	Overestimates are easily made. The technique is data-intensive	Part of the recreational value of a site is reflected in the amount of time and money that people spend while travelling to the site
	Hedonic pricing method	Reflection of service demand in the prices people pay for associated marketed goods	The method only captures people's willingness to pay for perceived benefits. Very data-intensive	Presence of communication in remote areas (via sat.) increases the attractiveness of area
3. Surveys	Contingent valuation method (CVM)	This method asks people how much they would be willing to pay (or accept as compensation) for specific services through questionnaires or interviews	There are various sources of bias in the interview techniques. Also there is controversy over whether people would actually pay the amounts they state in the interviews	It is often the only way to estimate non-use values. For example, a survey questionnaire might ask respondents to express their willingness to increase the level of water quality – which may rely on satellite data – in a stream, lake or river so that they might enjoy activities like swimming, boating or fishing
	Group valuation	Same as contingent valuation (CVM) but as an interactive group process	The bias in a group CVM is supposed to be less than in an individual CVM	See CVM above

Source: Adapted from De Groot, 2006 and OECD, 2006b.

In the satellite communications sector, the maritime markets are already well identified. They are quite diverse and include different types of users with different needs in terms of satellite services: the merchant ships (with fleet and ship management services, crew calling); the fishing community (messaging, position reporting services); the oil and gas offshore industry (large data transfers and positioning via satellites); government (data transfers *at sea*, encrypted services), passenger/cruise ship management (social calling, broadcast services); yachts/pleasure crafts (social calling, messaging, position reporting); and finally other services, including the international safety services, including equipment onboard required by the international Maritime Organisation (*e.g.* GMDSS, EPRIB) (see Chapter 4).

The maritime markets are dynamic ones, although with an adoption rate that is relatively slow compared to the land-mobile segment. According to NSR data (2007), the number of satellite units on maritime platforms will grow overall from 225 000 in 2005 to over 605 000 in 2012 and provide revenues of over USD 1 billion at the end of 2012 (Figure 6.2). Traditionally, the maritime markets mostly consist of narrowband products with voice and data terminals adapted to the rugged high-seas environment. With a little over 193 000 units in-service in 2005, the narrowband maritime market could grow to 381 000 units representing USD 470 million of revenues at the end of 2012. Direct broadcast satellite TV in the shipping and cruise industry has been on a high growth curve for the last few years and is a key product for crew comfort and passenger entertainment, growing from around 25 000 units in 2005 to possibly 200 000 in 2012 and with retail revenues in the latter year of over USD 221 million

Figure 6.2. **Global Satellite Maritime Units In Service, 2005-2012**

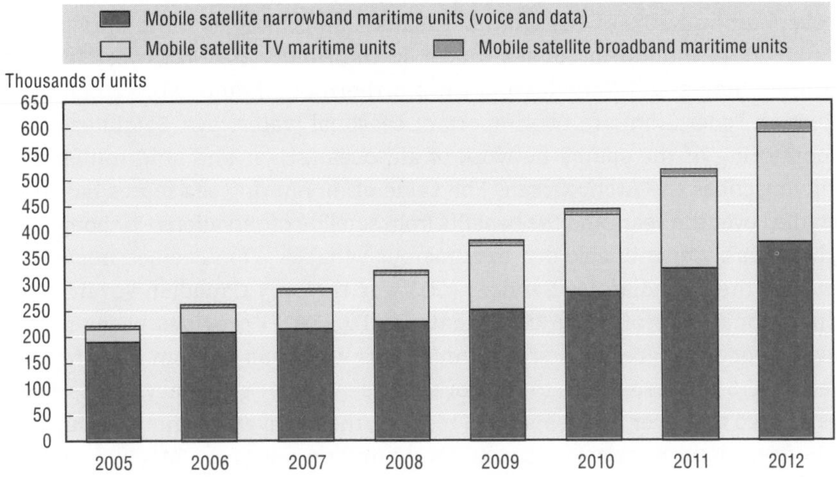

Source: Adapted from NSR, 2007.

(NSR, 2007). Broadband capabilities allowing Internet access are also becoming recognised as an important part of ship operations and the market is growing steadily. Consequently, a large number of firms are developing new products and services to target commercial merchant fleets (Thuraya, 2008).

Production factor – In economic theory, production factors are the resources employed to create goods and services. The factors can cover a large range of elements, such as labour, capital and land, but also technology. The use of space technology is seen in some economic sectors as a key factor in rising productivity or efficiency. A number of examples follow, based on particular experiences in fisheries, maritime transport, maritime zone control and weather forecasting.

Adoption of satellite navigation-related technologies in fishing fleets began in the mid-1980s, and general technology rollout and adoption began in the 1990s all over the world. The fisheries sector has mainly benefited in terms of reduced operating costs attributable to the use of GPS-enabled plotting systems. According to a recent Australian study (ACIL Tasman, 2008), some fishers stated that they saw their productivity rise by 50% or more as a result of GPS technology. Indeed, total Australian fishing industry output has increased by around 50% since the late 1980s (from around 180 000 tonnes in 1988/89 to around 270 000 tonnes in 2003-04). Other improvements in technology (boats, sonar scanning, nets, etc.) will also have contributed to this trend. The best available scientific evidence indicates that the fishing power of the fleet increased by around 12% due to the uptake of GPS and plotters. The cumulative addition to fishing output over time that can be conservatively attributed to the use of GPS plotters was estimated at 4.14% of output in 2007, equivalent to around AUD 88 million at 2007 prices (ACIL Tasman, 2008).

With regard to maritime transport, as seen in previous chapters, sea ice covers around 10% of the world's oceans – and a key satellite application comes from monitoring of sea routes. In the Arctic, the Northwest Passage (United States and Canada) and the Northern Sea Route (Norway and the Russian Federation) are two important seasonal waterways. Satellites allow monitoring of the sparse network of air, ocean, river and land routes that circumscribes the Arctic Ocean. The value of monitoring sea routes has been studied over the years and the benefits from satellite observations – although not always easily quantifiable due to numerous variables – are deemed important. In 2005 the Canadian Ice Services (CIS) was the only Canadian government operational user of RADARSAT-1 data. RADARSAT-1 provides observations over a wider geographical area in much less time than with an aircraft. As a result, CIS has been able to improve its operational efficiency. It has been estimated that over five years (1995 to 2000), the net average annual savings to CIS operations have been about CAN 7.7 million per year (CAN 38.5 million over five years), with the same per year benefits continuing up to and including the eighth year of operations for RADARSAT-1. The Canadian Coast

Guard (CCG), the largest direct customer of CIS products, has perhaps benefited most. The CCG Ice Operation Centres can provide improved routing information to commercial shipping, which allows for faster transit times. The shipping industry has also benefited from the accuracy of RADARSAT information to produce ice charts. The shipping companies believe that as a result of these ice charts, savings in their transit time through ice-infested waters are an estimated CAN 18 million a year. Other benefits included less damage to ships and reduced need for CCG escorts. The CCG has estimated dollar savings in both operating costs and transit time for those escorts to be between CAN 3.6 million and CAN 7 million a year, depending on the severity of ice conditions (CSA, 2005).

Box 6.3. Detection of oil pollution in the Mediterranean Sea

In the framework of its Monitoring Illicit Discharges from Vessel (MIDIV) programme, the Joint Research Centre (JRC) of the European Commission was tasked by the Italian Ministry of Environment to detect oil pollution in the Mediterranean Sea over the period 1999-2004 near the Italian coasts. Italy was then considering the creation of an environmental protection zone along some of its coasts. Using 18 947 radar images from archives (hence no aerial or ship validation) of the European satellites Envisat, ERS-1 and ERS-2, 9 299 possible oil spills were detected. It was estimated that a rather low spatial resolution (cheaper than higher resolution imagery) of about 200 metres was sufficient for statistical investigations of marine oil pollution. Most of the spills were away from the coasts, indicating deliberate intention to avoid possible legal actions in territorial waters. An interesting finding from the research links the decreasing number of oil spills in some areas to recent decisions by France to create an environmental protection zone, with increased aerial surveillance (JRC, 2007).

In 2006, an operational satellite-based oil slick detection service integrating SAR data from Envisat and the Canadian Radarsat satellite was set up for all European waters under the European Maritime Safety Agency (EMSA). The service, named *CleanSeaNet* has a long-term objective to be integrated into national and regional response chains, so as to strengthen operational pollution response for accidental and deliberate discharges from ships and assist coastal states in locating and identifying polluters in areas under their jurisdiction (EMSA, 2008). The service provides notification of a pollution event within 20 to 30 minutes of the satellite overpass. By integrating the SAR oil slick information with vessel information, it becomes possible to identify potentially responsible vessels. In parallel, a number of complementary or sometimes competing national and regional demonstration projects are under way.

Source: JRC, 2007; EMSA, 2008.

To offer another illustration of efficiency derived from satellites, there are a number of studies concerning maritime zone monitoring, for illegal fishing and ship-sourced pollution. A direct link between the usage of satellite imagery and the decrease in polluters or illegal fishers in a given area is not clear. However, the improved ship detection over large geographic zones enabled by integrating satellite imagery with other tools (*e.g.* aerial patrols) is indeed beginning to help deter illegal fishing and oil spills.

In the case of illegal fishing, in 2004 France set up a ground receiving station on the Kerguelen Island (South Indian Ocean), to monitor its exclusive economic zone. All Envisat and Radarsat-1 satellite overpasses there are acquired, processed and correlated with the French fishing Vessel Monitoring System (authorised fishing ships in the area are required to carry onboard a detector) – and followed up by ship patrol – to protect the local stocks from illegal fishing (Greidanus, 2005). Since then, it has been estimated that the surveillance system has cut the number of illegal fishing incursions in the vicinity by nine-tenths by late 2005, and no illegal incursion was detected in 2007 (French Assembly, 2008).

With regard to ship-sourced pollution, Canada has a challenging task when trying to monitor its enormous maritime area. The Canadian Exclusive Economic Zone (EEZ) extends 200 nautical miles offshore and contains over 5.5 million square kilometres, often bordering a complex crenulated coastline. However, according to data collected since 1993, the National Aerial Surveillance Programme has had important impacts on reducing pollution from passing ships in the Pacific Canadian EEZ (Serra-Sogas *et al.*, 2008). It is thought that the NASP is particularly effective as a deterrent because it involves regular flights over ships based at least in part on the indications provided by Radarsat satellite imagery to detect potential polluters (Figure 6.3).[3] The increasing effectiveness of radar imagery contributed to Norway's decision in 2003 to invest in the Canadian Radarsat programme. The objective was primarily to secure access to operational radar data for national oil spill detection, ship detection and ice monitoring services that had been developed by using data from ESA research satellites. In addition to the Canadian Exclusive Economic Zone example, the trend of decreasing oil spills was also found in parts of the Mediterranean Sea; there too it is linked to recent decisions by France to create an environmental protection zone with increased aerial surveillance and use of satellite imagery (EC /JRC, 2007).

Finally, an interesting example comes from the United Kingdom in terms of efficiency derived from satellite data for meteorological forecasts. The Meteorological Office and several Numerical Weather Prediction centres (NWP) have undertaken impact studies concerning data from polar-orbiting satellites. This is a complex question since the weather prediction products can be affected as much by improvements in modelling techniques as by improvements in initial

Figure 6.3. **Reduction in oil spills in the Canadian Pacific Exclusive Economic Zone (1993-2007)**

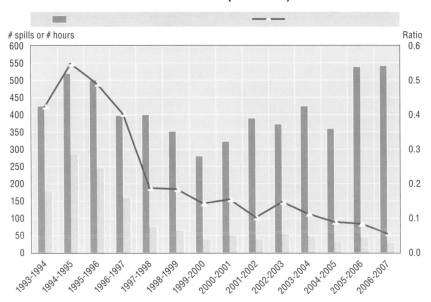

Source: Sogas *et al.*, 2008.

data. Nevertheless, the success of a forecast system can be measured in terms of accuracy through an index devised for the purpose (BNSC, 1998). A Met Office internal review of the situation early in 1998 showed that some 3.5% improvement in the index could be attributed to use of the raw satellite data stream operated by NOAA (at that time, Europe only operated satellites in geostationary orbit and relied on US low earth orbit polar satellites for data over sections of the Atlantic). Further improvements introduced during 1999 raised the index by about 5%, and impact studies showed that about 2% of that 5% could be attributed to better processing of NOAA data. The imagery was therefore still considered an essential forecasting aid. On the technical side, it was estimated that this development and provision of instruments had allowed the Met Office to develop a skill base, enabling it to act as an intelligent customer for procuring "best value for money" satellites through the various space programmes. Cost savings and improved performance through these activities was estimated to be many times the cost of retaining the team.

Indirect market valuation

When direct market valuation techniques are not sufficient (in terms of markets, efficiency, productivity), it is necessary to resort to more indirect means of assessing values. A variety of valuation techniques can be used to establish the (revealed) willingness to pay (WTP) or willingness to accept compensation (WTA) for the availability or loss of these services (De Groot, 2006). In particular for the space sector, these include new services that allow society to avoid costs that would have been incurred in the absence of those services. A prime example relates to flood management.

Flood control was the subject of very early cost-benefit studies of water resources (Brouwer, 2005); today, fairly sophisticated CBA procedures are used on a routine basis in most OECD countries. The focus is very much on probabilistic analysis of floods, the costs of their control and the damages avoided through that control. Damages are fairly easy to estimate; they range from property damages, impacts on wetlands and health risks to welfare losses arising from the "fear of floods". Many case studies have been conducted over the years as new space systems have come online; these assess the latter's role in flood management (*e.g.* high resolution satellites, increasing use of combined existing space systems), and the development of easier-to-use GIS systems and broadband telecommunications. Satellite monitoring generally provides large area coverage; it is relatively inexpensive once the satellite is in orbit if one can get the imagery at low cost (*e.g.* case by case agreements). However, it still requires calibration with ground-measured data. And although there are broad uses for remote sensing in delineating land use/land cover, its application in locating a very distinct water line with high accuracy is limited. Much of the problem is that vegetation shifts on the ground are not clearly evident in imagery (Smith *et al.*, 2004). Despite those limitations, the main benefits identified over the years include avoided costs of damages and improvement of operational efficiency (Table 6.6). In Italy the improvement of forecasting and early warning tools (including the use of space-based data) allowed a marked reduction in human losses from 1994 to 2000, according to the Italian Civil

Table 6.6. **General flood management benefits**

Better forecasting with damage costs avoided	Weather forecasting combined with hydrological models permits more advanced flood and flash flood warnings. Hence, – Reduced casualties and injuries (as well as consequences for public health). – Reduced economic damages (to property and economic activities) as well as environmental damages (especially relating to forest fires).
Improved operational efficiency	The response during the flood can be well targeted, satellite data providing the basis for extent mapping and real-time monitoring, assessing damage to infrastructure, meteorological assessments, evaluation of secondary disasters, resulting in other costs savings: – Reduced prevention costs (prevention plan elaboration, flood protection investments, forest maintenance). Quantification and assessment of the potential impacts were based on information and validation provided by the end-users directly involved in the projects. – Reduced anticipation costs (flood forecasting services, fire alert systems). – Reduced crisis management costs (rescue activities, firefighting, recovery).
Reconstruction	Satellite use speeds the reconstruction efforts and loss assessments (insurance).

Source: OECD, 2005 and Risk EOS, 2005.

Protection Agency (Soddu, 2006). The International Charter for Space and Major Disasters signed in October 2000 is another interesting case. It is a mechanism maintained by nine space agencies that agreed to supply their respective optical and radar imagery in times of natural disasters to requesting countries and international organisations. The Charter was activated 175 times in the last eight years. The latest activation was made in May 2008, following the earthquake in Sichuan province: China, a Charter member, used data as inputs to elaborate maps for rescuers.

Table 6.7. **Actions and estimated benefits of using space assets for flood management**

Location	Action	Benefits
Arles, France (2003)	Water pumping	During Arles flooding in the South of France (2003), 25 million m^3 of water were pumped by French civil protection. It took three weeks to get the city dry again. By comparison, in 2001 – before the International Charter for disaster was active and regular satellite observation was possible – it took three months for the cities and villages in the French *département* of the Somme to be dry again, because of the difficulty controlling the efficiency of water pumps on a large scale.
North Carolina, United States (2003)	Maps for flood insurance	North Carolina did an update of its flood insurance maps using remote sensing data. A cost-benefit analysis by the US Geological Survey mentioned that the state would gain USD 3.35 for each dollar spent on the programme, and lose USD 57 million each year that it did not have updated maps (NRC, 2003). As an added benefit, the maps could be used for community planning and other purposes.
California, United States (1997-98)	Evacuations based on weather forecast	Improved forecasts of the 1997-98 El Niño events are estimated to have saved California residents on the order of USD 1 billion compared to the costs of a similar event in 1982-83, which was not forecast.

Source: OECD / IFP.

The survey approach (contingent and group valuation)

One way to estimate a potential service's value is to ask users directly what they would do in a hypothetical scenario, via contingent valuation techniques. For example, a survey questionnaire might ask respondents to express their degree of willingness to pay for a service or an activity. Although methodological advances have been made with this approach, it is not without its problems (Box 6.4). Over the past few years, the related method of "contingent choice" – asking respondents whether or not they would pay a predetermined amount – has also gained popularity, since it eliminates some of the weaknesses of contingent valuation. Another approach involves group deliberation, or "group valuation". Here, small groups of people are brought together in a moderated forum to deliberate about the economic value of specific services.

Box 6.4. **Background on the contingent valuation method**

The contingent valuation method is perhaps the most widely used stated preference method or survey-based technique. Particularly from the 1990s onwards, the method has been applied extensively to the valuation of environmental impacts in both developed and developing countries. The range of environmental issues addressed is wide: water quality, outdoor recreation, species preservation, forest protection, air quality, visibility, waste management, sanitation improvements, biodiversity, health impacts, natural resource damage and environmental risk reductions, to list but a few.

Although still controversial, this direct survey approach to estimating individual or household demand for non-market goods has been gaining increased acceptance among both academics and policy makers as a versatile and comprehensive methodology for benefit estimation. This acceptance was in large part based on the conclusions of the special panel appointed by the US National Oceanic and Atmospheric Administration (NOAA) in 1993 (Arrow et al., 1993) following the Exxon Valdez oil spill in Alaska in 1989. The panel concluded that, subject to a number of recommendations, contingent valuation studies could produce estimates reliable enough to be used in a (US) judicial process of natural resource damage assessment. It is now over a decade since the NOAA deliberations, and recent important developments have included the publication of official guidelines for using stated preference research to inform UK public policy and state-of-the-art guidance on most aspects of non-market (environmental) valuation for the United States. Those developments have not been restricted to application of these guidelines and guidance in the field of environmental economics. There has also been important cross-fertilisation with health economics and, more recently, cultural economics. Most promisingly, much more is now known about the particular circumstances in which stated preference methods will work well – in terms of resulting in valid and reliable findings – and where problems can be expected. Such findings have had an important bearing on evolving best practice in the design of contingent valuation questionnaires.

Source: OECD, 2006.

In the space sector, those survey approaches are now used regularly to test the interest of potential users of future space systems and services. One defining factor for all of them is that the expected benefits usually concern the setting up of entire information systems, in which satellite data or signals are crucial but only one component in the long value chain. Follow the results of studies concerning: potential transit-time savings enabled by upcoming meteorological systems; cost avoidances in flood management; oil pollution monitoring benefits derived from satellite navigation and earth observation data; and potential benefits derived from better snow information.

In his report entitled *Benefits of NPOESS for Commercial Ship Routing: Transit-Time Savings*, Kite-Powell (2000) estimated the average time saved for container ships from routing with and without the future US National Polar orbiting Operational Environmental Satellite System (NPOESS) data (the system was then programmed for a 2007 launch). For Atlantic transits, the average time saved without NPOESS data is estimated to be 4 hours per transit; with NPOESS data it is estimated to be 7 hours per transit (a gain of 3 hours). For Pacific transits, the average savings without NPOESS data is 12 hours and with NPOESS data 21 hours (a gain of 9 hours). These savings of 3 and 9 hours are attributed to NPOESS working in concert with more traditional observations from the Polar-orbiting Operational Environmental Satellite (POES), the Defense Meteorological Satellite Program (DMSP), ground-based Doppler radar systems, airborne sounders and radar aircraft. Those time savings translate into an expected average annual benefit to ship routing from NPOESS data (in the two decades following the launch) of about USD 95 million per year. Because of the US share of world trade, perhaps 20% of the total benefit – some USD 20 million per year – could be realised by consumers in the United States.

Concerning flood management, a number of studies have focused on the future capabilities of upcoming systems and expected cost avoidances. The Risk-EOS study (2005) provides the results for a fully European risk information system, operating from 2012. Taking into account methodological constraints, the estimation mentioned of flood damages avoided has to be seen as very conservative. This is due to the fact that the economic costs avoided of flash flood damages are underestimated: it is not always easy to differentiate between flash floods and plain floods in available statistics.

Table 6.8. **Maximum benefits that can be expected from the European Risk-EOS services for floods**

Flood risk	Maximum expected benefits from Risk EOS (in 2020)	Benefits in 2012		Benefits in 2020	
		Value (million euros)	%	Value (million euros)	%
Flood preparedness and prevention	• 0.1% flood protection and operating and maintenance	1.1	0.6	7.4	0.8
Flood damages	• 2% of flood economic damages	70.8	36.2	351.6	38.5
Flash floods casualties	• 20% on flash flood casualties • 2% on plain flood casualties • 2% flood consequences on public health	27.25	13.9	118.7	13
Flood crisis management	• 2% flood rescue activities • 2% flood recovery costs	1.5	0.8	7.3	0.8

Source: Risk EOS, 2005.

With regard to oil pollution monitoring from space, in the GSE *Real-time Ocean Services for Environment and Security* (ROSES) study, a number of scenarios/ options have been identified to give an indication of the potential and prospective economic benefits a full detection system for oil spill could provide in Europe (Whitelaw *et al.*, 2004). In Option 1, only basic services are provided; global routine and on-request local services are offered, but with no advanced modelling capabilities. An Option 2 provides a full product set including all the types of service, while Option 3 extends the oil spill monitoring services to address marine primary productivity and sea level and climate change monitoring applications as well. The results are summarised in Table 6.9.

Table 6.9. **Potential societal benefits from an oil spill detection system (ROSES study)**

	% impact totals	Resulting benefit	Rationale
Option 1	Equivalent of 1.5% cleanup bill x 3 for discharge volumes	EUR 6.48M	Improved cleanup operations, % impact on the cleanup budgets averaged annually. Provision for two major events also included.
	Equivalent of 1.5% cleanup bill x 3 for discharge volumes	EUR 8.26M	Better detection has an impact on the levels of operational discharge through deterrence. Reductions in routine discharge with 50% increase after automatic identification system (or AIS, ship detection using satellite navigation signals) on stream in 2008. Same approach used as for cleanups of major accidents but scaled further by ratio of regular discharges (37K tons p.a.) to major spills (29K tons p.a.).
Option 2	2% of cleanup bill x 3 for discharge volumes	EUR 8.64M	As Option 1, but with increased percentage due to modelling contribution.
	Equivalent of 2% cleanup bill x 3 for discharge volumes	EUR 11.02M	Main advantage here is modelling of the effects of major spills. Reductions in routine discharge with 50% increase with AIS on stream.
Option 3	2.25% of cleanup bill	EUR 9.72M	As Option 1, but with further increased percentage due to more sophisticated modelling contribution (as with downstream).
	Equivalent of 2.25% cleanup bill x 3 for discharge volumes	EUR 12.4M	Main advantage here is modelling of the effects of major spills. Reductions in routine discharge with 50% increase after AIS on stream. Some additions for inputs to broader ocean circulation studies.

Note: The multiplication by a factor 3 (x 3) comes from an estimated ratio of social impacts to cleanup costs of about 3 to 1.
Source: Whitelaw *et al.*, 2004.

Other studies point to similar conclusions. For example, a possible 1% increase in oil spill containment and cleanup efficiency in the New England region would yield a savings of USD 7.5 million over ten years, and nearly USD 100 million for the entire United States over that same period (Adams *et al.*, 2000). As new quasi-operational systems are put in place using space-based imagery (in Canada and Italy but also in many other countries), with surveillance mechanisms added (*e.g.* deterrent aircraft patrols), one may be able to obtain better cost-efficiency measurement of the overall information and surveillance system.

With regard to future satellite navigation systems, NAUPLIOS (2002-04) was a pilot project of the 5th Research Framework Programme of the European Union, and managed by the EU Directorate General for Energy and Transport. The project's objective was to demonstrate the added value of the GALILEO positioning and search and rescue services for maritime transportation of goods and hazardous materials. One lesson learned concerns the role of satellite communications in the overall information system. Satellite communication faces two main drawbacks in comparison with VHF – costs and time delivery delays. It also, however, provides the great advantage of confidentiality in data exchanges. As to costs and benefits, it was found that use of GALILEO positioning and search and rescue services would also bring benefits in terms of avoided costs (*e.g.* oil spills and ecosystem degradations). In particular, total shipping accident costs leading to oil spills and other polluting could decrease – depending on scenarios with different levels of implementation – by 1% (1st scenario), 5% (2nd scenario) and even 10% (3rd scenario) (Table 6.10).

Finally, a 2004 report for NOAA on the value of snow information examines the potential benefits of setting up a dedicated monitoring system (Adams *et al.*, 2004). Snow imposes both economic costs and benefits on society, and these effects are significant in size relative to other weather phenomena. There is also value in using current information on snow coverage, snow pack water potential and other snow metrics in human decision making. Finally, improvements in the accuracy and lead time of snow metrics such as snow coverage, depth and water content could in particular improve individual and public decision making. Although the authors noted that a comprehensive costs/benefit study was not possible due to the lack of information (especially the costs of snow information services), several sector-level estimates of the economic impacts (benefits and costs) of snow and related winter events were provided (Table 6.11). The main conclusion of the study was that the value to society of a snow monitoring and reporting system would likely be substantially greater than the costs of this system (see also Stewart, Pielke and Nath, 2004).

Other promising valuation techniques (real options and portfolio)

Several methodologies come from the finance sector that are rather promising for a number of applications: the real options and portfolio methodologies. They are briefly discussed below.

Table 6.10. **NAUPLIOS project: Costs and benefits of GALILEO's added value**

Search and Rescue services for maritime transportation of goods and hazardous materials, 2004

Present value (2008)	Costs and benefits of GALILEO's added value in European waters (EUR 1 million, prices 2004)		
Costs :			
● Investment costs	**102.00**		
– Shipping industry	100.63		
– National Maritime Coordination Centres (NMCC)	0.06		
– Maritime authorities/ National Disaster Coordinating Council (NDCC)	1.31		
● Replacement, maintenance and exploitation costs	**21.36**		
– Shipping industry	20.28		
– NMCC	0.04		
– Maritime authorities/NDCC	1.03		
Total costs:	*123.36*		
	Scenario 1	**Scenario 2**	**Scenario 3**
Benefits:			
Total decrease in accident costs	1%	5%	10%
– Less costs resulting from oil spills due to accidents	45.3	226.4	452.7
– Less oil spills from illegal oil releases	21.3	106.4	212.8
– Increased safety resulting in lower casualties	20.6	103.2	206.4
– Decrease in number of dead animals (birds, seals, etc.)	+	++	+++
Total benefits:	**87.2** and +	**436.0** and ++	**871.9** and +++
Net present value (benefits minus costs)	−36.16 and +	312.64 and ++	748.54 and +++
Benefit/cost ratio (benefits/costs)	0.71	3.53	7.07

Note: National Maritime Coordination Centres (NMCC) co-ordinate in most countries the civilian use of maritime patrol and surveillance assets. In the first scenario it is assumed that the total accident costs will decrease by 1%, in the second scenario by 5% and in the third by 10%.
Source: Maréchal et al., 2004; ECORYS Transport, 2004.

The real options methodologies seek to uncover and quantify a project's embedded options or critical decision points. Adapting real options may yield interesting economic information about complex programmes, especially in organisations characterised by large capital investments, along with much uncertainty and flexibility (Teach, 2003). Although not used for space applications, recent experiences from the oil and gas, mining, pharmaceuticals and biotechnology industries indicate that this type of analysis provides valuable results (Archer and Ghasemzadeh, 2007). Companies in those industries also have plenty of the market or R&D data needed to make confident assumptions about uncertainties. Plus, they have the sort of engineering-oriented corporate culture that is not averse to using complex mathematical modelling (Reach, 2003). With real options, a number of constraints are

Table 6.11. **Examples of the economic impacts (benefits/costs) of snow and snow events in the United States**

(2004 dollars)

Economic benefits of snow	Winter tourism	Exceeds USD 8 billion/year in New England and Rocky Mountains
	Cold water fishing	Exceeds USD 2.3 billion/year in New England
	Snow pack water storage	Up to USD 348 billion/year in western United States
Economic costs of snow	Snow removal	Exceeds USD 2 billion/year for United States
	Road closures that cause lost retail trade, wages and tax revenue	Exceeds USD 10 billion/day for closures in eastern United States
	Flight delays	USD 3.2 billion annually for US carriers
	Damage to utilities	Up to USD 2 billion per event
	Flooding from snowmelt	USD 4.3 billion for 1997 floods
	Cost to agriculture and timber from frost and ice	Up to USD 1.6 billion per ice storm

Source: Adams *et al.*, 2004.

identified that are relatively useful when looking at space projects and derived applications:

● Budget constraints with fixed project costs.

● Logistical constraint, with mutually exclusive projects and other rigid interdependencies, such as follow-up projects.

● Positioning constraints help ensure that the composition of the portfolio is aligned with strategic requirements (*e.g.* starting a minimum number of projects in different technological or geographical areas).

● Threshold constraints to help ensure that the performance of the portfolio and its constituent projects fulfil minimum requirements (*e.g.* the aggregate net present value may have to exceed a minimum acceptable level).

Value remains in the eye of the beholder

A space-related service may bring substantial benefits or not, depending on a person's position in the value chain. For a direct end-user, the space application could be providing a solution, such as enhanced productivity. For a technician, it is piece of commodity equipment. To a financial officer, a space application may represent an overhead that needs to prove its worth. To a retailer, it may represent revenue.

A key variable in most CBA analysis is the choice of a "benefits denominator". Selected either as a preliminary educated guess (*e.g.* satellite data bring X% of benefits) or as a calculated average of the estimates provided by experts (*e.g.* PricewaterhouseCoopers' GMES study, 2006), or using a dedicated group contingent valuation method, the final number often remains contentious.

It is also very rarely directly attributable solely to satellite data, but instead most likely relates to a full information system or to derived services from this information system. As an example, the RISK-EOS study (2005) mentions that 20% of the flash flood casualties in Europe could be avoided thanks to the development of an information system using satellite-based data; such a system would also improve by 1% the efficiency and profitability of offshore production, resulting in gains in the range of EUR 340-840 million per year.

A great deal of literature considers how to assess the value of geospatial information in general, and types of weather information in particular. As mentioned by many users over the years, the value of earth observation, particularly for water management, comes mainly from providing evidence-based decision-making capabilities; it reduces uncertainty, although figures cannot always express that value. This is a problem that has plagued space-based remote-sensing activities over the past decades, since the taxpayers and their representatives are ultimately interested only in the final "outputs" of these activities.[4] According to Macauley (2004), space remote-sensing activities have not yet benefited from rigorous and consistent application of information valuation methodologies. The "value of information" approach is used to assess the marginal benefit of improved information due to better co-ordination. Filling information gaps largely depends on:

1. How uncertain decision makers are.

2. What is at stake as an outcome of their decisions.

3. How much it will cost to use the information to make decisions.

4. The price of the next-best substitute for the information.

In other words, users – actual or potential – of derived information from space-based data may place a certain value on it, which depends on their willingness to pay for such information. That willingness is a measure of what economists call "social surplus": the value of the information in excess of the costs of acquiring it. When such value accrues to businesses, it is referred to as "producer surplus"; when it accrues to individual users, it is called "consumer surplus". Many studies point to the rather low value of information, especially if the probability of an event is either very unlikely or very likely, or if the actions that must be taken to avert its effects are minimal.

In addition, being better informed does not always translate into action, and inaction may or may not have socio-economic consequences. This has an impact on the benefits expected to be derived from space systems. With regard to the value of weather information for example, the World Meteorological Organisation (WMO) noted that the distinction needed to be made between weather-sensitive activities and weather-information-sensitive activities. In weather-sensitive activity for example, coconut trees may be affected by severe weather, such as tropical cyclones – but there may not be much that can be done to save crops. In weather-information-sensitive activities, there is normally more scope for considering actions that could have significant economic implication – *e.g.* harvesting the matured rice crop days before the predicted passage of an incoming tropical cyclone (WMO, 2003).

In the larger context, there is a growing literature on attempts to assess the costs and benefits of possible alternative courses of action for decision makers. In the OECD, a research programme on the "costs of inaction" by policy makers is ongoing, led by the OECD Environment Directorate. While considerable work has been undertaken on the costs of implementing policy in specific areas, in many cases there is inadequate understanding of the consequences of inaction; those consequences may accrue exponentially in the distant future in a number of environment-related areas. The OECD is pursuing different strands of work on this issue in three areas: biodiversity and ecosystem services; climate change; and health impacts from pollution.

The European Environment Agency also looked at the "impacts of actions or inactions" of governments in responding to "early warnings" of hazards over the past hundred years (EEA, 2002). The report, entitled *Late Lessons from Early Warnings: The Precautionary Principle 1896-2000,* focused on lessons that could be learned from past histories to minimise possible future impacts of agents that may turn out to be harmful, and to do so without stifling innovation or compromising science. Looking at government reactions in a dozen case studies over the past decades (*e.g.* radiation in the early 1900s, sulphur emissions in the mid-1980s, "mad cow" disease in the 1990s), the report drew some interesting if sobering conclusions. There seems to be "no credible way of reducing the pros and cons of alternative courses of action to a single figure, economic or otherwise, not least because of the problem of comparing incommensurables and because the pros and cons are unlikely to be spread evenly across all interest groups."

The infrastructure approach

Large amounts of space-derived data and signals have become essential elements in the efforts to monitor and manage climate change. They have also provided useful inputs to a number of public good-oriented and commercial sectors, such as water management and maritime transportation.

However, the previous sections of this report suggest that alone, cost-benefit analysis of selected space applications does not appear to offer a satisfactory basis for decision making. On the cost side, it is clearly very difficult to allocate correctly the proportion of costs related to specific operational functions. Space activities remain an intensive R&D sector with long lead times, and satellites generally carry a multifunctional array of sensors. On the benefits side, it is difficult if not impossible to single out and measure the contribution provided by the satellites, as data and signals need to be integrated with other content and equipment to be useful.

Given these obstacles, other approaches to assessing the utility of space applications need to be explored. One such approach is to consider components of the satellite infrastructure as a public good-type infrastructure and to compare them with terrestrial infrastructures. This chapter will review some basic options for delivering a space-based infrastructure, and then examine the economic role of infrastructures in general. It will then draw parallels with investment in selected infrastructures.

Reviewing options for delivering a space-based infrastructure

Historically, space-based systems have been built through public R&D programmes designed to develop new technologies, conduct research in diverse scientific fields, and provide innovative capabilities. Since the 1980s, a number of space systems and their derived applications have proved profitable investments for private actors, particularly in the telecommunications sector. Meanwhile improved space and terrestrial technical capabilities – particularly in computing power and data processing – have led to increased use of satellite data and satellite signals in numerous applications by both public and private actors.

That is where a number of space programmes present an interesting paradox, as mentioned previously: they are considered and funded as R&D programmes but in many cases act as key infrastructures delivering unique public and private services. The term "space infrastructure" is defined as encompassing all space systems, whether public or private, that can be used to deliver space-based services. Both space and underlying ground segments are included. As mentioned in OECD (2005), there are two complementary and interlinked space-based infrastructures. The first one is in essence the "front office", i.e. "user-oriented" and designed to provide information-related services including communications, navigation and earth observation to governments and society at large. The second is the essential enabling "back office", i.e. the space transport, satellite manufacturing and servicing infrastructure. The following paragraphs will mainly focus on the "front office", particularly for earth observation.

There are several options for providing an infrastructure, from outsourcing to devolution to public or private actors (Table 6.12). As stated in OECD (2005), the role of government remains key in shaping space activities, because policy makers determine the rules of the game under which space actors – notably private ones – operate. Typically, governments are major users of infrastructure, whether it is public infrastructure to deliver services to citizens or private infrastructure as an input in their activities. In most cases, public services are financed by general taxes on the population at large, and provided free of charge or at marginal cost.

Table 6.12. **Outsourcing and devolution models for the provision
of infrastructure**

OUTSOURCING: Government retains overall responsibility for the provision of infrastructure, but selectively pays private companies to undertake specific operational tasks over limited periods, based on contractual arrangements:

1. Simple contracting out	The most basic level, this involves tendering out discrete activities, such as road works or tolling management, on a case-by-case basis.
2. Design-build arrangements	A further step involves the transfer of responsibility for designing and building infrastructure, as a single package, to a private partner.
3 Public-private partnerships (PPPs)	Transfer of extensive responsibility for the designing, building, operation, maintenance and/or financing of infrastructure, as well as associated risks, to private partners over long periods, after which the project is transferred back to government.

DEVOLUTION: Transfer of responsibility for the provision of infrastructure to entities that exist specifically for that purpose. To a greater or lesser degree, the decision-making processes within these organisations are not under the direct control of elected officials. Different models of devolution include, with increasing degrees of independence:

1. Government agencies	Public bodies that report directly to government ministries, but which typically have a more limited set of responsibilities and a higher degree of leeway with regard to operational decisions than a ministry would have. Agencies can be established both for the delivery of works and to manage funds dedicated to infrastructure.
2. State-owned companies	Companies that are organised under private company legislation and whose management is largely independent in its decision making, but which are subject to government control by way of ownership.
3. Mixed companies	Companies in which the government maintains an important ownership stake, but where there is also private ownership.
4. Private, not-for-profit organisations	Private entities that reinvest net revenues in the infrastructure asset, with management responsible before a board made up of stakeholders, which could include government.
5. 100% private owner-operators	Situations in which the infrastructure asset is the property of a private company, which therefore assumes responsibility for all aspects of its provision, based on commercial principles.

Source: Adapted from OECD, 2007c.

By going further and drawing parallels with other infrastructures, government must strike a delicate balance in devolving or outsourcing space-based infrastructure: that between the pursuit of new efficiencies and the need to oversee the maintenance and development of key public assets. Private financing of infrastructures often does not generate "new money", since ultimately most infrastructure must be paid for by some combination of users and taxpayers (OECD, 2007c). One exception is the mostly privately owned satellite telecommunications infrastructure, which benefits from a large profitable retail market (*e.g.* television broadcasting, maritime markets for mobile satellites services). But even in that sector, development of future capabilities relies on public funding of R&D programmes with long lead times (such as ESA's Artemis and Artes telecommunications programmes).

In the case of devolution, operational agencies may be run as purely public bodies financed by the state. [Examples include the European Organisation for the Exploitation of Meteorological Satellites (EUMETSAT) in Europe and the National Oceanic and Atmospheric Administration (NOAA) in the United States for meteorological satellites.] They may also be run on a "commercial" basis, generating a substantial share (if not all) of their revenue from the sale of services. (An example here is the Antrix Corporation Limited, the commercial arm of India's Department of Space.)

In the context of tackling climate change, one important lesson learned from the previous chapters is that space-related agencies are the cornerstones for key programme choices: working with scientists, operational users of different fields and industry, they contribute to the development of major R&D and operational systems. However, as the role of space-based infrastructures (particularly earth observation) is increasing, so should the role of R&D and operational agencies, with adequate funding in their respective lines of work.

Economic role of infrastructure

The economic impact of infrastructure has been the subject of much debate since at least the 1980s, with the discussion focusing on both the direction and magnitude of effects (OECD, 2006a; 2007d). While it is possible to establish a link between infrastructure development and economic development, it is difficult to ascertain the direction of causation: does infrastructure contribute to economic development, or *vice versa*? Moreover, there was considerable scepticism about the initial estimates of productivity gains stemming from investment in public infrastructure.

Over the past few years, however – with improved data, new methodological approaches and refinements to models – there has been much less controversy surrounding the benefits derived from large-scale infrastructures. The patterns of underinvestment in infrastructure in some countries may have something to do with the difficulties governments experience in estimating the overall long-term effects of infrastructure on the economy. Making the "right" decision regarding infrastructure development is often difficult because of the public good nature of the benefits (how much is enough, who should benefit). Moreover, the broader impact of infrastructure is clearly conditional on how efficiently it is used. Poorly managed or poorly conceived infrastructure may well generate less return. What is important to note is that the returns on infrastructure investment take time to materialise, and may take different forms.

A review of the more recent literature suggests that public infrastructure has a positive productive effect on the economy, but that the size of the effect is not as large as that estimated by earlier studies. Based on samples of several OECD countries and broken down according to economic sectors, the efficiency impacts of large-scale infrastructure tend to be positive – but relatively modest – in almost all sectors. There is general agreement that road construction for example does on the whole produce economy-wide gains, but estimates vary. However, it can be said that investment in road infrastructure in OECD countries generates macroeconomic productivity effects equivalent to a rate of return of between 5% and 8%. This is certainly modest. But it has to be said that in developing countries with inadequate road networks, economy-wide productivity gains from additional road construction tend to be significantly higher. As with roads, estimates of the benefits of improving water and sanitation infrastructure vary considerably, but they are thought to be very high in the poorer developing countries.

The productive impact of an infrastructure depends not only on the magnitude of the investment, the project's design and efficient management, but also on the nature of the investment and its integration into an existing set of infrastructures, i.e. how it improves the network. Thus, first infrastructures only have limited impact on public and private sector productivity since their effect is primarily local. The addition of new infrastructures to create a network, however, allows considerable productivity gains by extending the use of existing infrastructures. Subsequently, when the network is largely completed, the addition of new infrastructures once again has only limited (if any) impact on private sector productivity.

From this point of view, the space-based earth observation (EO) infrastructure presents some interesting parallels. The existing systems are largely funded nationally or regionally, like other large infrastructure projects (highways, rail), and their overall impacts increase as the network of systems expands. As in the case of interconnected road networks (i.e. linking more users), complementary EO systems provide specific added yields, such as a greater diversity of data, back-up capacity, and improved sustainability of the overall infrastructure. Too large an expansion of the infrastructure could create fewer benefits and even some loss of efficiency (duplications). However, this is not yet the case for EO systems, although the growth of national programmes in many countries (more than 16 with EO systems in 2006) may accelerate the trend.

As previously mentioned, increased productivity is often derived from *infrastructure expansion (network)* and *interconnections* with other infrastructures. In the case of satellites, the notion of network is well understood; it is demonstrated by efforts made at the international level by space countries to inform other countries about their national developments and co-ordinate with them (*e.g.* CEOS and GEO). In the case of meteorology, the co-ordinating role of the World Meteorological Organisation is essential, even if nations may in some cases limit full international accessibility to their national system's weather data. The necessary interconnections with other types of infrastructure include links with *in situ* systems, which contribute unique data complementary to space-based data, and telecommunications (*e.g.* terrestrial and satellite broadband), which provide the needed links to scientific and operational users.

Figure 6.4. **Overview of general impacts derived from setting up a space-based infrastructure**

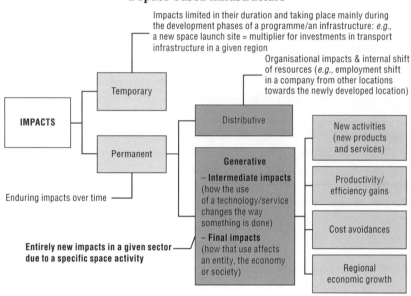

In the case of data coming from specific space systems, the question seems more crucial. The nature of earth observation systems is close to that of the traditional public good (Box 6.5). The basic characteristic of a public good is that its benefits are neither excludable (there is no way of charging for them) nor rival (one person's consumption does not reduce another's).

A number of space agencies have set up specific data policies so as to allow better access to scientific users. In April 2008 for example, the US Geological Survey announced it would gradually make all archived Landsat satellites' imagery available on the Internet for free (USGS, 2008). Today, many differences in data availability remain between countries, and different restrictions may apply (for security purposes, special licences). In addition, several commercial earth observation companies operate satellites and sell the data they collect. They are few though, and their investors and customers are often governmental bodies (Table 6.13).

A balanced approach between commercial data products – particularly the value-adding sector which facilitates end-users' access to satellite data (Wensink, 2008) – and the innate public good nature of some of the data needs to be found, possibly following the model of meteorological data. The benefits from earth observation are indeed largely the provision of knowledge, data and information that are generally considered a public good. Other typical examples of public goods are defence, and (uncongested) non-toll roads.

Box 6.5. **Definition of public good**

A *public good* is a commodity, measure, fact or service:

- Which can be consumed by one person without diminishing the amount available for consumption by another person (non-rivalry).
- Which is available at zero or negligible marginal cost to a large or unlimited number of consumers (non-exclusiveness).
- Which does not become unusable to any consumer now or in the future (sustainability). The degree of non-exclusiveness determines the public good's degree of purity.

Other definitions:

- An *international public good* (IPG) is a public good which provides benefits that cross the national borders of the producing country.
- A *regional public good (RPG)* is an international public good which provides spillover benefits to countries in the neighbourhood of the producing country, a region smaller than the rest of the world.
- A *global public good (GPG)* is an international public good which, while not necessarily to the same extent, benefits consumers all over the world.

Source: Adapted from Reisen *et al.*, 2004.

Table 6.13. **Selected commercial satellite operators in the earth observation sector**

Firms	Satellites	Status
Spot Image (FRA)	<ul><li>3 optical satellites (Spot 3, 4 and 5)</li><li>2010: Pleiades constellation (2 optical satellite with very high resolution 50 cm over 20x20 km)</li><li>2012: future high resolution sensor to ensure SPOT 5 service continuity</li></ul>	Mixed public-private funding (public private partnerships with CNES, DLR, etc.) EADS Astrium Services is the majority holder of both Infoterra and Spot Image
Infoterra Group (DEU, FRA, GBR)	<ul><li>1 high resolution radar satellite (TerraSAR-X, 1m over 10x10 km)</li><li>2009: 1 new radar satellite TanDEM-X (TerraSAR-X add-on for digital elevation measurements)</li></ul>	
RapidEye (DEU)	<ul><li>5 small radar satellite constellation, to be launched in summer 2008</li></ul>	
DigitalGlobe (USA)	<ul><li>2 optical satellites (Quick Bird and WorldView-1, allowing up to 50 cm resolution at nadir with a swath of 17.6 km)</li><li>2009: 1 new optical satellite (WorldView-2)</li></ul>	US privately-held companies, benefiting from specific US Department of Defense contracts for development and imagery (*i.e.* Clearview, Nextview)
GeoEye (USA)	<ul><li>1 high resolution optical satellite IKONOS (1 metre at nadir with a swath of 11.3 km)</li><li>1 new optical satellite (GeoEye-1), to be launched in summer 2008</li></ul>	
ImageSat (Netherlands Antilles)	<ul><li>2 high resolution optical satellites (EROS A and B, 1.9 metres at nadir with a swath of 14 km)</li></ul>	Netherlands Antilles company, with Israel Aircraft Industries Ltd. (IAI) as majority holder

Source: OECD/FP.

Comparative approach with terrestrial infrastructures

Arguably a key question for policy makers should be: are investments in a given space infrastructure to help meet climate change challenges inadequate, compared with investment in terrestrial infrastructures in other areas? To put the space-based earth observation infrastructure into a comparative perspective, parallels are drawn here with road and rail infrastructures, electricity infrastructures, and telecommunications (OECD, 2006; OECD, 2007b), even though some of those do not fit neatly into the public good category.

- *Road infrastructure* – Current stocks worldwide are estimated at over USD 5 trillion. Annual infrastructure investment is put at an annual USD 220 billion, equivalent to about 4.5% of world assets. Approximately 60% of that goes to maintenance and replacement, and 40% to net additions to road networks. In the coming decades the share of maintenance and replacement is expected to increase (up to as much as 80%) as stocks age and increasingly fewer new roads are built (especially in the more developed countries).

- *Rail infrastructure* – Assets worldwide are valued at some USD 630 billion. Annual infrastructure spending is estimated at nearly USD 50 billion, equivalent to around 8% of total assets.

- *Electricity transmission and distribution* – Assets worldwide are valued at some USD 3 trillion. Annual infrastructure spending is estimated at nearly USD 130 billion, equivalent to 4-5% of total assets.

- *Terrestrial telecommunications* – While this cannot be really called a public good, it is nonetheless an interesting infrastructure for comparative purposes given the ICT basis it shares with space technologies. Global stocks of telecommunications equipment (fixed and mobile) were valued at around USD 3 200 billion in 2005, and annual infrastructure investment at around USD 650 billion (splitting roughly 40/60 into new build and maintenance investment). Hence, yearly investments are equivalent to around 20% of the value of current assets, significantly higher than for roads, rail and electricity.

Table 6.14. **Estimated annual world infrastructure expenditure (additions and renewal) for selected sectors, 2005**

Type of infrastructure	Stock (USD)	Annual investment (USD)
Road	6 trillion	220 billion
Telecoms	3.2 trillion	650 billion
Rail	630 billion	50 billion

Source: OECD, 2006a.

Broadly speaking, the level of global investment going to terrestrial infrastructures over the next 10-15 years or so would appear to be increasing slightly in absolute terms but slowing as a share of GDP. This is very much consistent with the theory that as economies become more developed and networks more complete, the rate of new infrastructure added to the system declines, and the emphasis shifts from new build to maintenance. Table 6.15 provides an indication of the orders of magnitude of the infrastructure investment requirements worldwide over the next decades.

Table 6.15. **Estimated average annual world infrastructure expenditure (additions and renewal) for selected sectors, 2000-20, in USD billion**

Type of infrastructure	2000-10	2010-20
Road	220	245
Rail	49	54
Telecoms[1]	654	646
Electricity[2]	127	180
Water[1, 3]	576	772

1. Estimates apply to the years 2005, 2015 and 2025.
2. Transmission and distribution only.
3. Only OECD countries, the Russian Federation, China, India and Brazil are considered here.

In the case of the space-based earth observation infrastructure, additions and maintenance (i.e. fleet renewal, expanded services) and research and development (for new instruments and technologies) are the main – often overlapping – cost drivers. A clear separation between operational systems and R&D satellites is still not easily made, since the two types of satellites are complementary. Based on published observation requirements and already approved missions, a worldwide investment of roughly USD 38-40 billion, averaging USD 1.5 billion to a little over USD 3 billion a year, seems necessary for additions and maintenance in the next decade (2008-20).

Table 6.16. **Estimated annual investments (maintenance, replacement, expansion) in earth observation (2006, 2005, 2004)**

	Annual investments (in billion USD and as % of total in-orbit assets at end-2006)	
2006	3.2	15
2005	1.1	6
2004	1.6	10

Source: OECD/IFP.

The above suggests that levels of investment (annual spend as a proportion of total assets) in earth observation infrastructure are higher than those for roads and electricity network infrastructures, but close to those one would expect for expenditures on rail networks and telecommunications. Roads and rail infrastructures can have very long life spans – hundreds of years in some cases. This has to be borne in mind when comparing them with space-based assets, which have very much shorter operational lives – generally between five and ten years. Given this shorter life span, the conclusion one is tempted to draw is that the rate of replacement and expansion of the earth observation infrastructure (USD 1.6 billion in 2004, 6% of total; USD 3.2 billion in 2006, 15% of total) is relatively low compared with that for terrestrial infrastructures.

Comparative approach with terrestrial data and information infrastructure

The closest parallel with a weather information system as an information-intensive infrastructure is perhaps a country's statistical agency. The role of economic indicators – as imperfect as they may be sometimes (*e.g.* because of methodological issues) – is essential in today's modern societies. They serve as markers of the health and performance of an economy, even providing when necessary alert mechanisms.

While assets are hard to quantify, data do exist on national annual budgets to support operations. For most OECD countries, these operational budgets range between 0.02% and 0.05% of national GDP. These appear to be modest investments for statistical infrastructures that after all underpin economic and social performance.

Interesting parallels can be found when comparing national statistical offices with the budget of operational weather agencies, although they have clear differences in terms of missions (particularly for R&D).

Two examples are provided: Eumetsat and NOAA's information satellite services. The European Organisation for the Exploitation of Meteorological Satellites (Eumetsat)'s mission is to deliver weather and climate-related satellite data, images and products to the National Meteorological Services that are members of the Organisation. Eumetsat has an annual budget of around EUR 150 million to EUR 200 million (EUR 168 million in 2008), with regular peaks when developing and procuring new satellites (Eumetsat, 2008). But on top of this Eumetsat budget, major investments in European meteorological satellites come from ESA and individual EU member states, up to around 65% for the development of the space segment of a given programme (*e.g.* national R&D efforts for a specific instrument that will be carried on board the European meteorological satellites).

Table 6.17. **Budgets of various OECD national statistical offices in USD and as a % of national GDP**

	Year	Office/bureau	Estim. Budget (USD million)	% of national GDP
Australia	2005/06	Australian Bureau of Statistics[1]	243	0.039
Norway	2005	Statistics Norway[2]	77	0.031
Sweden	2005	Statistics Sweden[3]	59	0.018
Canada	2005/06	Statistics Canada[4]	528	0.052
France	2005	INSEE[5]	415	0.022
Germany	2005	Federal Statistical Office[6]	183	0.007
Italy	2003	ISTAT[7]	282	0.021
United States	2005	Census Bureau[8]	765	0.007
United Kingdom	2005/06	Office for National Statistics[9]	389	0.020

1. Revenues from government appropriations and other sources – Australian Bureau of Statistics Annual Report.
2. Total government appropriations, revenues and refunds – *Statistics Norway Annual Report*.
3. Sweden has a turnover of SEK 911 million but budget appropriations of SEK 440 million – *Statistics Sweden*.
4. Gross budgetary main estimates – Treasury Board of Canada.
5. INSEE, *Rapport d'activités 2005*.
6. Germany's expenditure for all levels of government was EUR 4.302 billion.
7. ISTAT annual budget.
8. Funding for discretionary appropriations, permanent appropriations and budget authority – FY 2007 *Budget in Brief*.
9. Gross administration budget – Office for National Statistics Annual Report and Accounts.
Source: OECD/IFP.

NOAA plays a national meteorological services role for the United States. It is in charge of setting up requirement and procuring meteorological satellites, in co-operation with NASA. NOAA's budget is around USD 1 billion for its satellite activities (NOAA, 2008). Again, funding for US meteorological satellites does not come solely from NOAA; there are other sources, particularly the Department of Defense. For example, the future National Polar-orbiting Operational Environmental Satellite System, or NPOESS programme, is funded equally by the Department of Commerce which oversees NOAA and the Department of Defense Air Force annual appropriations.

A risk management approach to investment in space-based infrastructures

To help reduce uncertainties related to climate change and so improve decision making, a risk management approach applicable to investment in space-based infrastructures is explored here. This section begins by introducing some notions about risk management and uncertainty, and then draws some parallels with weather information. Finally, it envisages systematic space-based climate monitoring as a compelling approach to reduce uncertainty.

Introduction to risk management and uncertainty

The risks to human life and economic assets stemming from the effects of population growth, economic growth, globalisation and climate change are substantial, and difficult to predict. By the time those effects are felt, they may well be irreversible. Indeed, some of the natural hazards facing society are so great in magnitude (hurricanes, earthquakes, tsunamis, droughts of continental proportions, pandemics), and the potential economic impacts so severe, that some catastrophe risks would appear to be uninsurable. A mixture of private insurance, financial market instruments and state funding would be needed to counter the dramatic losses. By way of illustration, a repetition of the 1923 Tokyo earthquake would today inflict losses in the order of one-third of Japan's current GDP (Zajdenweber, 2000).

Risk management comprises an array of methods and techniques that estimate the likelihood and consequences of undesired events, using either qualitative or quantitative methods (An example of the latter is the uncertainty or "insufficient weight of the evidence" coefficient developed by Keynes.) Those tools provide valuable if not perfect decision aids in the face of uncertainty. As shown almost a century ago by Knight (1921), one way to deal with large-scale uncertainty is to insure against a range of outcomes.

Knight's concept has been studied extensively by those developing the economic theory of insurance (see Dionne, 2000; Eeckhoudt, 2005). Insurance follows a rather basic risk transfer mechanism: premiums are paid as a mark-up over potential losses. A typical problem for insurance companies is the possibility of incurring losses above premium income, especially if independent losses add up too quickly and so produce a high aggregate loss. A possible solution is reinsurance, which pools the portfolio of insurance companies. Risks linked to climate uncertainty, however, might become too large to bear with traditional instruments. Thus, new insurance investment vehicles have emerged in the past decade; they are presented in the next subsection.

Parallels with weather information

Weather information has always been useful for reducing uncertainty and making decisions – from planting crops since antiquity to decisions on the most cost-effective energy outputs when the temperature rises in summer. Over the past decades, improvements in the ability to forecast weather have clearly had an important impact on society.

Better forecasts than ever – Today's four-day weather forecast is as accurate as two-day forecasts were 20 years ago, and the accuracy of forecasts of large-scale weather patterns in both hemispheres has been increasing since 1980. The traditional error in a three-day forecast of the landfall position of hurricanes has been reduced from about 337 kilometres in 1985 to about

177 kilometres in 2004 (SSB, 2005). This ability was instrumental in warning residents in the United States during the series of severe hurricanes in 2005. Figure 6.5, adapted by SSB (2005), shows the monthly moving average of the correlation between forecast and observed weather features for 3-day, 5-day, and 7-day forecasts. A perfect forecast is 100% (vertical axis). The accuracy of forecasts of large-scale weather patterns in both hemispheres has been increasing steadily since 1980. The southern hemisphere forecast (bottom curves), which in 1980 was significantly worse than the northern hemisphere forecast (top curves), has caught up in accuracy in recent years. This dramatic improvement has been due largely to more and better global satellite data.

Figure 6.5. **Correlations between forecast and observed weather features for 3-day, 5-day and 7-day forecasts**

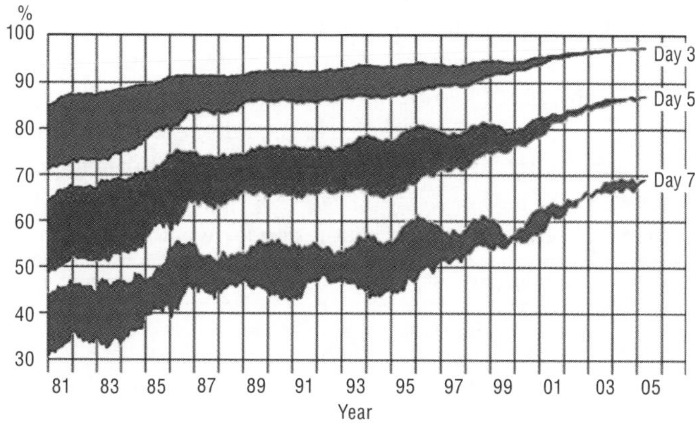

Source: SSB, 2005, adapted from Simmons and Hollingsworth, 2002.

Benefits derived from weather forecasts – There are different methodologies for estimating the economic benefits of meteorological activities. They are mostly based on value of information methodologies, such as normative or prescriptive decision making; descriptive behavioural response studies (user surveys, regression models) including contingent valuation; computable general equilibrium or economy-wide models; and "market prices" to measure the benefits of private-good meteorological services (WMO, 2003; Gunasekera, 2003). Most studies focus on prescriptive models of decision making by individual businesses in particular in the agriculture sector. Stated choice studies have also been conducted to estimate the public's willingness to pay for improved weather products, and how much people value the weather services currently provided.

As mentioned by WMO, though the economic benefit estimates currently available are still limited (in particular marginal, *i.e.* incremental benefits), they are substantial. During a major WMO 1994 conference, certain estimates given in a number of papers are still cited today in diverse presentations (Abedayo, 2006). A typical factor for the ratio of economic benefits to a National Meteorological Hydrological Service (NMHS) budget may fall in the range of 5-10. As a crude approximation, and given that the global budgets of NMHSs in 1994 was in the region of USD 4 billion, it was concluded that the global economic benefits were therefore in the range of USD 20-40 billion, although clearly this was only a broad indicative estimate (WMO, 2003).

In the United States, a 2002 report for NOAA found that the average value of weather forecast information relative to total federal spending produces an annual benefit-cost ratio of 4.4 for US households, or net national benefits of USD 8.8 billion a year (Lazo and Chestnut, 2002). This estimate does not include derived benefits in agriculture, transportation or construction, or benefits to households in other countries that rely on weather information from the United States. It was also found that in a typical hurricane season, NOAA's forecasts, warnings, and the associated emergency responses result in a USD 3 billion savings (Willoughby, 2001). Two-thirds of this savings, USD 2 billion, is attributed to the reduction in hurricane-related deaths, and one-third, USD 1 billion, to a reduction in property-related damage because of actions taken to prepare.

Table 6.18. **Selected benefit cost ratios for weather information**

Benefit ratio	Overall benefits	Source
5-10 (NMHS)	Ratio of 5-10:1 economic benefit to a National Meteorological Hydrological Service budget in 1994. Translates to global economic benefits in the USD 20-40 billion range.	WMO, 2003
4.4 (US households)	Annual benefit cost ratio of 4.4 for US households, or net national benefits of USD 8.8 billion a year (without derived benefits in agriculture, transportation, etc.).	Lazo and Chestnut, 2002, Report for NOAA
35 to 40	Using different methods, the meteorological service benefit-cost ratio in China ranges from 35 to 40; includes both public meteorological services and meteorological services for various economic sectors (survey of 1 279 experts from all types of sectors).	Zhang and Wang, 2003
0.57 (Personal users) 9 (Construction sector) 15 (Agriculture)	In 1980, Eurostat undertook cost-benefit analyses of the Meteosat programme, estimating the direct benefits in comparison with the budgetary costs, in a series of weather-sensitive industries. According to these estimates, the benefit-cost ratio was 0.57:1 for personal users and society, 9:1 for the construction industry and 15:1 for agriculture.	Cohendet and Lebeau, 1987

In Europe, Eurostat undertook a prospective cost-benefit analysis of the Meteosat programme in 1980; the goal was to estimate direct benefits in comparison to budgetary costs in a series of weather-sensitive industries.

According to these estimates, the benefit-cost ratio was 0.57 for personal users and society, 9 for the construction industry and 15 for agriculture (Cohendet and Lebeau, 1987). More recently, a detailed survey among Meteosat users in the United Kingdom also estimated these benefits (Morel de Westgaver and Robinson, 2000) and extrapolated them to the rest of the participating countries. That study found that better weather forecasting has led to substantial benefits in the air transportation industry: about EUR 11 million per year. Benefits to the agricultural industry were about EUR 30 million, thanks to the rational use of pesticides. Overall accumulated benefits for the general public reached around EUR 2.75 trillion over a period of 20 years. Meteosat also allowed a reduction in costs in the fields of transportation, energy, fishing, agriculture and the manufacturing industry of up to EUR 2.5 trillion (estimates based on declared preferences expressed in a survey).

One recurring problem, identified in previous sections, is the diversity and at times complexity of existing methods for the evaluation of benefits derived from meteorological services. This makes international and economic sector comparisons indeed challenging.

Economic importance of weather information: the insurance market – Cost-benefit analysis provides valuable data, but another way to look at the economic importance of weather information is to study insurance markets.

With regard to insuring uncertainty concerning weather, complementary mechanisms have been developed over the past fifty years. They include premiums set up by insurance companies, reinsurance, and the use of markets for risk transfer particularly via catastrophe ("cat") bonds. Weather is a major determinant of earnings performance for entire economic sectors (*e.g.* utilities). For decades, financial products based on weather outcomes have been available in most OECD countries, to transfer weather risk to counterparties in a better position to manage it (Campbell and Diebold, 2005). There are two main weather risk insurance packages: catastrophe risks and weather risks (Table 6.19). Historically, weather hedges have usually been bought by energy companies and agriculture actors. It has been estimated for instance that up to 80% of agricultural losses are linked to weather conditions (WMO, 2002).

To offer another illustration of the usefulness of weather insurance, the US Department of Transportation estimates that weather-related delays cost passengers USD 10 billion in lost time and productivity each year in the United States alone. As shown in Table 6.20, June is historically a high-delay month in the United States, and two-thirds of those delays can be related to weather events. An analysis of American Airlines' weather-related losses in 2004, which topped USD 97 million, showed that extreme weather alone was responsible for USD 1 million in losses, whereas other events averaged USD 200 000 (Anselmo, 2007).

Table 6.19. **Financial instruments for weather insurance**

Catastrophe risks (e.g. floods, hurricanes, earthquakes)	Weather risks (e.g. temperature and rainfall fluctuations)
• Low-frequency high-severity risks (unlike motor vehicle risks for example). • Probabilities of occurrence and damage not precisely computed. Need much more data than for high frequency risks. • Variance of loss is high. Premium setting is difficult. • Capital requirements to ensure solvency are large. • Premiums can be high (as high as seven times the expected losses – the actuarially fair level). • Premiums can change drastically with an event – suggesting that probabilities of extreme events are not well established and therefore revised with any new information.	• Different types of coverage/severity depending on the sector (temperature and rainfall fluctuations with agriculture crops, storms with airline industries) • Often the insurance package is for systemic aggregate risk but leaves out individual-specific risk • System encouraged by the World Bank in developing countries, particularly for farmers

Table 6.20. **US airline delays accountable to weather**

	June 2003	June 2004	June 2005	June 2006	June 2007
Number of flight delays that month	27 284	44 282	39 128	32 172	41 447
Percentage of all delays that month	74.5%	77%	78%	70%	72.3%

Source: US Department of Transportation, quoted in Anselmo, 2007.

With regard to climate change, the number of industries interested in the risk management approach is growing; there are new insurance customers in transport, retail, tourism and construction. New weather risk management insurance instruments, such as catastrophe bonds or "cat bonds", are also being put in place by the private sector to help industries limit their financial exposure to climate and, as shown in Figure 6.6, transfer the economic risks of catastrophes onto the international stock market. Record-setting years are becoming commonplace; 2007 was by far the most active year in the brief history of catastrophe bond issuances, following Hurricane Katrina and the 2004-05 hurricane season.

As of early 2008, the total catastrophe limit outstanding – i.e. the maximum amount of insurance that can be paid for a covered loss – amounts to USD 169 billion. Among the available instruments the cat bond market is becoming significant, representing 12% of estimated property limits outstanding in the United States. That country is the world's "peak" exposure zone, where theoretically capital markets have the greatest ability to absorb large losses (GC Securities, 2008; see Figure 6.7).

Another tool is the recently developed industry loss warranty (ILW), a financial mechanism that covers large-scale losses from events where the industry-wide insured loss exceeds some pre-agreed threshold (i.e. the loss is at

Figure 6.6. **Annual catastrophe bond transaction volume in the United States, 1997-2007**

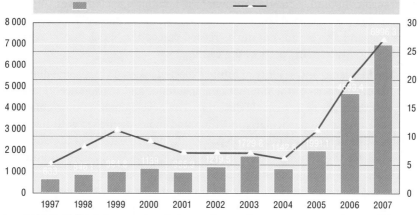

Source: GC Securities, 2008.

Figure 6.7. **Limits outstanding for world and the United States in early 2008**

A. Global property limits: USD 169 billion

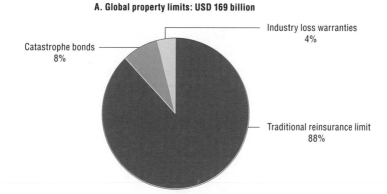

Catastrophe bonds
8%

Industry loss warranties
4%

Traditional reinsurance limit
88%

B. US only property limits: USD 81 billion

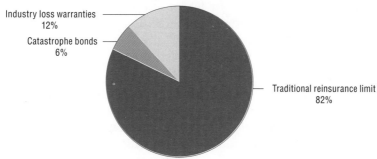

Industry loss warranties
12%

Catastrophe bonds
6%

Traditional reinsurance limit
82%

Source: GC Securities, 2008.

industry level rather than at company level because of a large-scale disaster).[5] Although the premiums for this type of coverage are expensive, this trend of risk transfers to markets could continue based on the expected excess demand for coverage (i.e. increased bond issuance) compared to supply (i.e. the pool of investors), and also because of the increased perception of risk due to extreme weather activity in many parts of the world.

The frequency and severity of future expected losses do cast some doubt on the financial capacity of the international insurance and reinsurance industries to absorb the costs of large-scale disasters. The development of new emerging insurance markets – particularly in China and India, countries with histories of large-scale natural hazards – could potentially place greater pressure on global markets over the coming decades. The high risk of accumulation and the difficulties in spreading catastrophic risks, geographically as well as over time, are among the main problems that insurance and reinsurance companies are facing in this field. Capital markets may provide additional sources of capacity, as shown in this section, but their role should not be overestimated at this stage (OECD, 2004).

Reducing uncertainty with systematic space-based climate monitoring

Based on the analysis conducted so far, decision makers have several options when tackling uncertainties linked to climate change:

- They can decide not to take any specific step to reduce the uncertainties identified at the beginning of the report. However, this may come at cost. Costs of inaction have already been evaluated by several organisations, all which point to important potential impacts in terms of economic growth and loss in GDP (OECD, 2008b). Moreover, those estimates may moreover well be underestimates since extreme events and non-market impacts are still often not included.

- They can rely on transfers of the risks of extreme weather events to individuals and markets, via insurance, reinsurance and capital markets (catastrophic bonds, for example). As seen in Chapter 1, insured natural peril losses have been on the rise for the past 20 years, mainly because of the increased value concentration in high-risk populated areas, higher vulnerabilities and widening insurance coverage. But another recurring factor is the notable rise in extreme weather events and the consequent effect on the ability to absorb costs mentioned above.

- They can decide to reduce uncertainties by developing the right tools to make better-informed decisions. Among those tools, space technologies – particularly earth observation – provide unique capabilities, as shown in Chapter 4. The data becoming available internationally on climate and the environment are expected to increase. However, there are still questions on how much other key data might be *missing* if new national and international missions to replace current systems are not launched soon. This is especially true for some climate-related satellite measurements.

Improving satellite infrastructure is indeed but one of the avenues that need to be pursued to generate reliable data on climate change and its possible impacts. As shown in Figure 6.8, a better scientific understanding of the phenomena described throughout this report will only be gained by taking into account the interrelationships between better observing systems, improved analyses and modelling, physical understanding, and clearer documentation of observed patterns in climate.

Figure 6.8. **Interrelationships between inputs and components leading to better understanding of weather extremes**

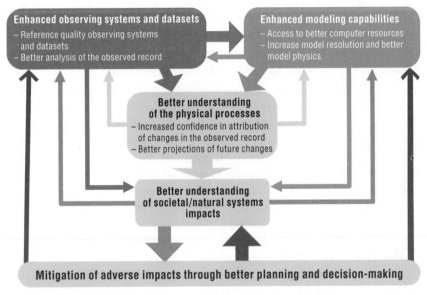

Source: Based on CCSP, 2008.

As part of this climate monitoring system, a number of space technologies are making very valuable contributions to the management of the earth's resources – mapping hazards, deepening knowledge of natural phenomena, and so on. Despite the difficulty in assessing quantitatively the socio-economic benefits derived from space-based infrastructures, unique capabilities have been identified that have positive impacts on both scientific research and operational monitoring of fundamental variables and possible tipping points of climate change (*e.g.* the Arctic situation). In terms of helping to manage major risks, their contribution is multifaceted.

Thus, space-based infrastructure and its derived applications can help:

● Improve understanding of the risks.

● Reduce uncertainty.

● Reduce vulnerabilities.

● Strengthen prevention.

● Enhance the conditions for mitigation.

For space-based climate monitoring to develop as a sustainable routine activity, several modifications to existing structures must be made. As mentioned in previous chapters, space-based observations of climate variables have been relying a great deal on R&D missions, in addition to operational meteorological systems. A large number of climate variables have required – and still require – the development of dedicated new sensors for scientific research on climate processes (WMO, 2007). For many years, numerous climate observations have been made possible by R&D missions. R&D space agencies have successfully made, and continue to make, a key contribution to climate monitoring. But today this approach raises some questions.

The GCOS Climate Monitoring Principles (GCMPs) require long-term continuity of measurements, which is not the primary objective of R&D programmes. In addition to R&D activities, which are obviously required to further progress in science and technology, what is needed is recognition of climate observations as operational programmes. Already, major national and international efforts have been made in terms of meteorology. Many satellites from different countries form the space-based Global Observing System (GOS) that the World Meteorological Organisation co-ordinates, and from which key weather information is available daily to citizens and private actors. The GOS framework is based on voluntary commitments by WMO members. Building on experiences from the meteorology world, systematic climate monitoring may become an essential tool for governments to hedge the risks associated with climate change and unsustainable resources management.

Policy makers can create for themselves and the populations they serve the opportunities to be warned in advance and to better manage potential impacts. As a possible way ahead for earth observation infrastructure in particular, more attention should be given to building on major decades-long national and international efforts to develop and sustain operational satellite meteorology. For climate monitoring to develop fully as a routine activity with long-term continuity of measurements, and with induced socio-economic benefits, institutional work-sharing and adequate funding will become increasingly necessary for agencies responsible for satellite R&D activities, and the operational weather agencies will necessarily inherit new climate-related tasks.

Space infrastructure – particularly earth observation – needs to be considered as a strategic asset in an infrastructure portfolio approach, when decision makers are forced to consider their options for improved risk management.

Notes

1. The European Organisation for the Exploitation of Meteorological Satellites (Eumetsat) is an intergovernmental organisation that establishes and maintains operational meteorological satellites for 19 European States. Eumetsat is currently operating Meteosat-6, –7 and –8 over Europe and Africa, and Meteosat-5 over the Indian Ocean. The data, product and services from those satellites make a significant contribution to weather forecasting and to the monitoring of the global climate.

2. NASA was to incur costs for operating the TRMM satellite through 2007 even if the mission had been terminated in December 2004 (as planned), since it takes time for a spacecraft to drift down to an appropriate altitude for controlled re-entry.

3. The Canadian National Aerial Surveillance Programme is operated by Transport Canada, and is the principal surveillance mechanism for monitoring and enforcing ship compliance with the international MARPOL regulations (i.e. oil pollution at sea) covering the Atlantic, Pacific and Arctic EEZs, as well as the St. Lawrence – Great Lakes Seaway.

4. Measurable units of inputs include spacecraft, instruments, staffing and operations costs. Units of output – that is, the value of the information gleaned from data, beyond merely counting bytes of data or the number of earth observation or weather "products" supplied – are more difficult to measure.

5. Examples of what ILWs could cover: a winter freeze with industry-wide insured loss in North America exceeding USD 20 billion; an earthquake with industry-wide insured property loss exceeding USD 35 billion anywhere in the world.

Bibliography

ACIL Tasman (2008), "The Value of Spatial Information: The Impact of Modern Spatial Information Technologies On the Australian Economy", Report prepared for the CRC for Spatial Information and ANZLIC, Spatial Information Council, Melbourne, Australia, March.

Adams, Richard et al. (2000), The Economics of Sustained Ocean Observation: Benefits and Rationale for Public Funding, National Oceanic and Atmospheric Administration and Office of Naval Research, Washington, DC.

Adams, Richard and Laurie L. Houston, Rodney F. Weiher (2004), The Value of Snow and Snow Information Services, Report prepared for NOAA's National Operational Hydrological Remote Sensing Center.

Amesse, Fernand and P. Cohendet (2001), "Technology Transfer Revisited from the Perspective of the Knowledge-based Economy", *Research Policy*, Vol. 30.

Anselmo, Joseph (2007), "Protection Money: New Venture Offers Hedges Against Increasingly Uncertain Weather", *Aviation Week and Space Technology*, 24 September.

Archer, Norman and Fereidoun Ghasemzadeh (2007), "A Decision Support System for Project Portfolio Selection", *International Journal of Technology Management*, Vol. 16, Nos. 1-2, 17 August.

Arrow, Kenneth *et al.* (1993), "Report of the NOAA Panel on Contingent Valuation", *Federal Register,* Vol. 58, No. 10, 15 January.

Bach, Laurent, Patrick Cohendet and Éric Schenk (2002), "Technological Transfers from the European Space Programmes: A Dynamic View and a Comparison with Other R&D Projects", *Journal of Technology Transfer*, Vol. 27, Issue 4, December.

Beka, Eric (2007), "La Belgique n'a que de bonnes raisons d'investir dans le spatial", *L'Écho*, 6 July.

BETA (1989), "Indirect Economic Effects of ESA Contracts on the Canadian Economy", Final Report.

BETA (1997), "Indirect Economic Effects of ESA Contracts on the Canadian Economy (Second Study, 1988-1997)", Final Report.

BNSC (British National Space Centre) (1998), *Evaluation of Funding for UK Civil Space Activity, Chapter 5: The Met Office Report*, London.

Brouwer, Roy, ed. (2005), *Cost Benefit Analysis and Water Resources Management*, Edward Elgar Publishing Limited, London.

Brown, Peter J. (2007), "MSS: New Services Changing the Market", *Via Satellite*, 1 November.

Campbell, Sean and Francis X. Diebold (2005), "Weather Forecasting for Weather Derivatives", *Journal of the American Statistical Association*, Vol. 100.

Capron, Henri and Didier Baudewyns (2007), "Impact des Établissements scientifiques fédéraux sur le développement économique, social et culturel de la Belgique: étude", Politique scientifique fédérale, SP1837, Brussels, March.

Capron, Henri, Didier Baudewyns and Marie Depelchin (2007), *Évaluation monétaire de certains avantages générés en Belgique par les ESF belges*, Note intermédiaire, DULBEA-CERT, ULB, Brussels.

CCSP (US Climate Change Science Program) (2008), *Weatherand Climate Extremes in a Changing Climate. Regions of Focus: North America, Hawaii, Caribbean, and US Pacific Islands*, US Climate Change Science Program, Synthesis and Assessment Product 3.3, June.

CNES and INSEE (2005), *Économie du spatial en Guyane, www.cnes.fr/web/3919-economie-du-spatial-en-guyane.php*, accessed December 2006.

Cohendet, P. and André Lebeau (eds.) (1987), *Les grands programmes civils*, Economica, June.

Comité de Bassin Rhône Méditerranée (2005), "La crue du Rhône de décembre 2003: quels enseignements", Colloque d'Arles, 3 December.

Cowell, R.G., R.J. Verrall and Y.K. Yoon (2007), "Modeling Operational Risk with Bayesian Networks", *Journal of Risk and Insurance*, December.

CSA (Canadian Space Agency) (2005), *Status Report on Major Crown Projects: Report on Departmental Performance 2004*, *www.espace.gc.ca/asc/eng/resources/publications/report_mcp-2005.asp?printer=1*, accessed November 2007.

De Groot, Rudolfand Mishka Stuip, Max Finlayson (2006), Valuing wetlands: Guidance for valuing the benefits derived from wetland ecosystem services, Ramsar Technical Report No. 3, CBD Technical Series No. 27.

Dionne, Georges, ed. (2000), *Handbook of Insurance*, Kluwer Academic Publishers.

ECORYS Transport (2004), "Evaluation of NAUPLIOS Initiative: Cost Benefit Analysis", Final Report, Prepared for the European Commission DG TREN, ECORYS Transport, Rotterdam, 27 April.

Eeckhoudt, Louis, Christian Gollier and Harris Schlesinger (2005), *Economic and Financial Decisions under Risk*, University Press, Princeton.

EMSA (European Maritime Safety Agency) (2008), EMSA Website: *http://cleanseanet.emsa.europa.eu*, accessed on 5 March.

Eurospace (1994), *Direct and Indirect Economic Impact of European Space Activities*, ESA Contract 11011, Paris.

FAA (Federal Aviation Administration) (2006), *The Economic Impact of Commercial Space Transportation on the US Economy: 2004*, February.

French Assembly (2008), "Rapport de la Commission des Affaires Étrangères sur Le Projet De Loi No. 443, autorisant l'approbation de l'accord entre le Gouvernement de la République française et le Gouvernement de l'Australie relatif à la coopération en matière d'application de la législation relative à la pêche dans les zones maritimes adjacentes aux Terres australes et antarctiques françaises, à l'île Heard et aux îles McDonald", Report No. 686, Paris.

GAO (Government Accountability Office) (2008a), *Space Acquisitions: DOD Is Making Progress to Rapidly Deliver Low Cost Space Capabilities, but Challenges Remain*, GAO-08-516, Report to the Subcommittee on Strategic Forces, Committee on Armed Services, US Senate, Washington DC, April.

GAO (2008b), *Space Acquisitions: Major Space Programs Still at Risk for Cost and Schedule Increases*, GAO-08-552T, Statement of Cristina T. Chaplain, Director, Acquisition and Sourcing Management, Testimony Before the Subcommittee on Strategic Forces, Committee on Armed Services, US Senate, Washington DC, March.

GC Securities (2008), *The Catastrophe Bond Market at Year-End 2007: The Market Goes Mainstream*, Sixth Edition, Guy Carpenter and Company, LLC, New York.

Glad, Stephen (2005), "Booz Allen Hamilton Uses Parametric Cost Estimating Tool to Move Navy's Satellite Communication Program to Next Level", *Cost Engineering Journal*, January.

Greidanus, H. (2005), "Assessing the Operationality of Ship Detection from Space", Proceedings of New Space Services for Maritime Users: The Impact of Satellite Technology on Maritime Legislation, UNESCO Headquarters, Paris, 21-23 February.

Gunasekera, Don (2003), "Measuring the Economic Value of Meteorological Information", *WMO Bulletin*, Vol. 52, No. 4.

House, P. Kyle *et al.* (2002), "Ancient Floods, Modern Hazards: Principles and Applications of Paleoflood Hydrology", *Water Science and Application Series*, Vol. 5.

Keynes, John Maynard (1921), *A Treatise on Probability*, Watchmaker Publishing; Unabridged edition, 5 May 2007.

Kite-Powell, Hauke L. (2002), *Benefits of NPOESS for Commercial Ship Routing – Transit Time Savings*, Marine Policy Center, Woods Hole Oceanographic Institution.

Kleinmuntz, Don N. (2007), "Resource Allocation Decisions" in Ward Edwards, Ralph F. Miles and Detlof Von Winterfeldt (eds.), *Advances in Decision Analysis 2007*, Cambridge University Press.

Knight, Frank H. (1921), *Risk, Uncertainty and Profit*, Dover Books on History, Political and Social Science, 2006 edition.

Lazo, J.K. and L.G. Chestnut (2002), *Economic Value of Current and Improved Weather Forecasts in the US Household Sector*, Report Prepared for the National Oceanic and Atmospheric Administration, Stratus Consulting Inc, Boulder, Colorado. Liesiö, Juuso, Ahti Salo and Pekka Mild (2008), "Robust Portfolio Modeling with Incomplete Cost Information and Project Interdependencies", *European Journal of Operational Research*, Vol. 190, Issue 3, 1 November, pp. 679-695.

Macauley, Molly K. (2004) "The Value of Information: A Background Paper on Measuring the Contribution of Space-Derived Earth Science Data to National Resource Management", Discussion Paper 05–26, Resources for the Future, prepared for NASA, May 2005.

Marechal, Jean (2004), "To All the Ships at Sea", *GPS World*, July.

Morel de Westgaver, Eric and Paul Robinson (2000), *Economic Benefits from ESA Programmes*, ESA publications, Paris.

Munich RE (2005), "Weather Catastrophes and Climate Change: Is There Still Hope for Us?", *Geo Risiko Forschung*.

NASA (National Aeronautics and Space Administration) (2002), *NASA cost estimating handbook*, NASA HQ, Washington DC, Spring.

NASA (2006), *NASA Facts: John C. Stennis Space Center*, FS-2006-05-00027-SSC, May.

Neumann, John von and Oskar Morgenstern (1944), *Theory of Games and Economic Behavior*, Princeton University Press.

NRC (National Research Council) (2003), "Using Remote Sensing in State and Local Government: Information for Management and Decision Making", Steering Committee on Space Applications and Commercialization, Washington DC.

NRC (2006), Assessment of the Benefits of Extending the Tropical Rainfall Measuring Mission: A Perspective from the Research and Operations Communities, Washington DC.

NSR (Northern Sky Research) (2007), Mobile Satellite Services, 3rd edition report, June.

OECD (2004), "Policy Issues in Insurance No. 8: Catastrophic Risks and Insurance", Proceedings of the OECD Conference on Catastrophic Risks and Insurance, 22-23 November, OECD, Paris.

OECD (2005), *Space 2030: Tackling Societies' Challenges*, OECD, Paris.

OECD (2006a), *Infrastructure to 2030 (Vol. 1): Telecom, Land Transport, Water and Electricity*, OECD, Paris.

OECD (2006b), Cost-Benefit Analysis and the Environment: Recent Developments, OECD, Paris.

OECD (2007a), *OECD E-Government Project: Benefits Realisation Management*, Report prepared for the 35th Session of the Public Governance Committee, 12 – 13 April 2007, Salle des Nations, Tour Europe, GOV/PGC/EGOV(2006)11/REV1, 29 March.

OECD (2007b), *Infrastructure to 2030 (Vol. 2): Mapping Policy for Electricity, Water and Transport*, OECD, Paris.

OECD (2007c), *Transport Infrastructure Investment: Options for Efficiency*, OECD Joint Transport Research Centre and the International Transport Forum, OECD, Paris.

OECD (2007d), *Transport Infrastructure Investment and Economic Productivity*, ECMT Round Tables No. 132, Paris.

OECD (2007e), *Assessing the Socio-Economic Impacts of Public R&D*, DSTI/STP/TIP(2007)18, OECD, Paris, 28 November.

OECD (2007f), *The Space Economy at a Glance*, OECD, Paris.

OECD (2008a), *OECD Environmental Outlook to 2030*, OECD, Paris.

OECD (2008b), *Economic Aspects of Adaptation to Climate Change: Costs, Benefits and Policy Instruments*, OECD, Paris.

PR Newswire (2008), "Maritime and Aeronautical Broadband Will Drive Growth in Mobile Satellite Services, New Report Claims", *PR Newswire*, 1 April.

Raghu, K. (2008), "Isro Plans Satellite Series for Mapping, Climate Monitoring", *Mint*, 15 January.

Reisen H. and M. Soto, T. Weithöner (2004), *Financing Global And Regional Public Goods Through ODA: Analysis And Evidence From the OECD Creditor Reporting System*, OECD DevelopmentCentre, Working Paper No. 232, DEV/DOC(2004)01, January.

Romer, P. (1990), "Endogenous technological change", *Journal of Political Economy*, 98, S71-S102.

Sankar, U. (2007), *The Economics of India's Space Programme: An Exploratory Analysis*, Oxford University Press, New Dehli.

Schumpeter, Joseph A. (1912), *The Theory Of Economic Development: An Inquiry Into Profits, Capital, Credit, Interest, and the Business Cycle*, Transaction Publishers, 1982 Edition, New York.

Serra-Sogas, Norma *et al.* (2008), "Visualization Of Spatial Patterns And Temporal Trends For Aerial Surveillance Of Illegal Oil Discharges In Western Canadian Marine Waters", *Marine Pollution Bulletin*, Volume 56, Issue 5, May.

Smith, Scott *et al.* (2004), "Assessment of the Use of Remote Sensing Techniques for Locating and Mapping Ordinary High Water Lines for Lakes Kissimmee and Hatchineha in Florida", *Surveying and Land Information Science*, American Congress on Surveying and Mapping, June.

Soddu, Pierluigi (2006), "Focus on Risk Matrix: The Italian Civil Protection System", Presentation to OECD, Rome, 11 October.

Solow, R. (1956), "A Contribution to the Theory of Economic Growth", *Quarterly Journal of Economics*, Vol. 70, pp. 65-94.

Space Studies Board (SSB) (2005), *Earth Science and Applications from Space: Urgent Needs and Opportunities to Serve the Nation*, Committee on Earth Science and Applications from Space, National Research Council, Washington DC.

Stewart, Thomas R., Roger Pielke and Radhika Nath (2004), "Understanding User Decision Making and the Value of Improved Precipitation Forecasts: Lessons from a Case Study", *Bulletin of the American Meteorological Society*, Vol. 85.

Teach, Edward (2003), "Will Real Options Take Root? Why Companies Have Been Slow to Adopt the Valuation Technique", CFO Magazine, 1 July.

Thuraya (2008), "Thuraya Targets Maritime Market With ThurayaMarine", Satellite Today, 2 June.

USGS (US Geological Survey) (2008), "Imagery for Everyone", USGS Press Release, 19 April.

Wensink, Han (2008), "From Data to Information: The Role of Value-Added Companies in Facilitating End-users' Access To Satellite Services", EARSC Workshop, Local and Regional Risk Management: Integrated Use of Satellite Information and Services, 10–11 March.

Whitelaw, Alan et al. (2004), "Real-time Ocean Services for Environment and Security (ROSES) Cost Benefit Analysis", Report prepared by ESYS for ESA, 29 October.

Willoughby, H, (2001), Costs and Benefits of Hurricane Forecasts, minutes of 55th Interdepartmental Hurricane Conference, 5-9 March 2001, Orlando, Florida.

WMO (World Meteorological Organisation) (2002), "Applications of Climate Forecasts for Agriculture", Proceedings of an Expert Group Meeting for Regional Association (Africa), 9-13 December, Banjul, Gambia, AGM-7, WCAC-1, WMO/TD-No. 1223.

WMO (2003), "Socio-economic Benefits of Meteorological and Hydrological Services", WMO Bulletin, Vol. 52, No. 4.

WMO (2006), Preventing and Mitigating Natural Disasters, WMO-No. 993.

WMO (2007), "Final Report: WMO Workshop on the Redesign and Optimization of the Space-based GOS", WMO Headquarters, Geneva, 21-22 June.

WMO/Global Climate Observing System (GCOS) (2008), "Vision of the Space-Based Global Observing System to 2025", Atmospheric Observation Panel For Climate, Fourteenth Session, Geneva, Switzerland, 21-25 April; Doc. 21a, WMO/IOC/UNEP/ICSU GCOS, 14 April.

Zajdenweber, Daniel (2000), Économie des extrêmes, Flammarion, Paris.

Zhang, Guocai and Haixiao Wang (2003), "Evaluating the Benefits of Meteorological Services in China", WMO Bulletin, Vol. 52, No. 4.

ISBN 978-92-64-05413-4
Space Technologies and Climate Change
Implications for Water Management, Marine Resources
and Maritime Transport
© OECD 2008

ANNEX A

The OECD Forum on Space Economics

Background

In February 2006 the Organisation for Economic Co-operation and Development (OECD) launched a Global Forum on Space Economics under the aegis of the International Futures Programme (IFP). This Forum helps agencies and governments better identify the statistical contours of the space sector and investigate its economic importance and impacts on the larger economy. Its work builds on the results and recommendations presented in the OECD publications *Space 2030: The Future of Space Applications* (2003) and *Space 2030: Tackling Society's Challenges* (2005).

Objective

The Forum aims to assist space-related agencies and governments by providing evidence-based analysis on space infrastructure (evaluating data and socio-economic indicators), so that the potential of that sector for the larger economy is more fully realised.

Participants

Participants in the Forum include the British National Space Centre (BNSC), Centre National d'Etudes Spatiales (CNES), Canadian Space Agency (CSA), European Space Agency (ESA), Italian Space Agency (ASI), National Aeronautics and Space Administration (NASA), National Oceanic and Atmospheric Administration (NOAA), Norwegian Space Centre (NRC), and the US Geological Survey (USGS). Other agencies and ministries from OECD and non-OECD countries can be observers and/or join. A companion Working Group, including the Group on Earth Observations (GEO) Secretariat, is open to interested parties and representatives of the private sector during selected meetings.

Activities

Three types of activities have been undertaken by the Forum so far:

1. Dedicated work on statistics and economic indicators, so as to contribute with other actors to the emergence of international comparative data on the space sector and its contribution to broader economic activity.

2. Work on horizontal case studies, which are meant to explore the wider economic and social dimensions of space applications (*e.g.* case study on satellites' contribution to water management).

3. An annual update of the state of the sector via the prism of the main OECD/IFP Space Project recommendations.

Selected Space Forum outputs

- *The Space Economy at a Glance* (2007): the first OECD statistical overview of the emerging space economy, including an innovative compilation of statistics on the space sector and its contributions to economic activity.

- *Climate Space Technologies and Climate Change: Implications for Water Management, Marine Resources and Maritime Transport* (2008): An OECD publication examining the socio-economic contributions that may be derived from the use of space technologies, with an extensive review of existing cost benefits methodologies.

- *Measuring the Space Economy: A Guide* (2008): An OECD Working Paper providing methodologies for indicators that could enable actors to build data series and comparable statistics.

- *Report on Monitoring the Space Project Recommendations* (2007 and 2008 reports): An annual state of the sector via the prism of OECD/IFP recommendations.

ANNEX B

Basics on Satellite Sensors

With space-based observations, the capacity to usefully measure water-related phenomena depends on the type and quality of measurements of the electromagnetic spectrum performed by remote sensing instruments. Those measurements depend on the general parameters of the satellite, which include 1) the field of view of the sensor, 2) the satellite's orbit, 3) the revisit frequency, 4) the spatial resolution and 5) the spectral bands used by the sensor.

1) *Field of view* – Currently, the higher a satellite's altitude, the larger the field of view it provides, but with less detail (resolution). Conversely, satellites in low earth orbit produce better resolution but a smaller field of view. Even if a satellite has a rather large swath width that allows it to view large areas, that does not mean it will be capable of collecting information for all areas. A satellite indeed can only acquire a certain amount of data; exactly how much depends on a range of factors – the number of ground stations, the size of the on board recorder if there is one and other technical limitations.

2) *Orbit* – The orbit of a satellite will dictate the location, time and frequency of observations. The two main types of satellite orbits are polar orbits and geosynchronous orbits:

● Polar orbit satellites fly in low earth orbits (generally 800 – 7 000 km) and acquire observations at nearly all latitudes, revolving around the earth 14 times a day. When in sun-synchronous orbit (a special polar orbit where the satellite orbital plane's rotation matches the rotation of the earth around the sun), the satellite passes over a point on earth at the same local solar time each day.

● Satellites in geosynchronous orbit (also known as geostationary) are at a higher altitude, around 36 000 km. Their orbits keep them synchronised with the earth's rotation; hence they appear to remain stationary over a fixed position on earth. They provide an almost hemispheric view and have an added advantage: frequent monitoring of events. For example, the NOAA

Geostationary Operational Environmental Satellite (GOES) provides visible and infrared images every 30 minutes with a spatial resolution of 1 to 4.

3) *Revisit* – The revisit of a region depends on the orbit of the satellite, the latitude of the observed area, and the field of view of the instrument carried on board the satellite (swath *vs.* measurement point). In general, the higher the sensors' resolution, the lower the revisit rate will be. At this time polar orbit satellites do not provide adequate coverage of processes that require hourly or daily monitoring. Flying several sensors in a constellation is a possible solution to obtain better revisit times, as well as to provide the instruments with pointing capabilities. The latter would allow them to increase the frequency of their observations, even when not flying over the desired area. The satellite sensors currently providing near-real-time capability to monitor the biogeochemical, optical and physical processes of coastal and open oceans are quite limited. They consist of the Sea Viewing Wide Field of View Sensor (SeaWIFS), the Moderate Resolution Imaging Spectroradiometer (MODIS, on NASA's Terra and Aqua satellites), the Medium Resolution Imaging Spectrometer (MERIS), Hyperion, and the Landsat Thematic Mapper (TM).

4) *Spatial resolution* – This is an important remote sensing characteristic. The higher the resolution, the better the detail can be detected on the earth's surface, at the expense of wider coverage.

- *Low and middle resolution* – Those systems can accurately measure large-scale phenomena such as weather patterns and changes on the earth's surface. They are the main programmes, which could be extended and expanded in most countries (*e.g.* IRS, Landsat).

- *High resolution* – More recent civilian high-resolution systems (1-5 m), with technologies inherited from military systems, provide much more detail but cover less area and have less multispectral capabilities.

5) *Spectral sensors* – Whether a satellite uses optical or radar technologies, the usefulness of data from an earth observation system is directly related to the spectral bands used for data acquisition. The following sensors increase the amount of information gathered in a single image. They offer the possibility of identifying objects and phenomena based on the spectral signature of the object observed:

- *Panchromatic (Pan)*: imagery from the visible and near-visible spectrum is collected and recorded as shades of grey (black and white pictures); it often boasts excellent ground resolution (*e.g.* the *Pan* instrument on Ikonos).

- *Multispectral*: a few spectral bands of light are recorded individually and then viewed either together to produce a true-colour image, or in selected combinations to produce a false-colour image (*e.g.* the High Resolution Visible IR HRVIR instrument on Spot).

- *Superspectral*: this configuration has many more spectral channels (typically > 10) than a multispectral sensor (*e.g.* the MODIS instruments on NASA's Terra and Aqua satellites).

- *Hyperspectral*: images composed of a hundred or more contiguous spectral bands, which allows collection of a lot more information concerning one specific area (*e.g.* the Hyperion instrument on NASA's Earth Observing-1 (EO-1) satellite).

The more channels there are on a sensor, the larger the electromagnetic spectrum visible will be. The number of spectral bands in imaging sensors has gradually been increasing over the past decades (*e.g.* from five channels in AVHRR to 36 channels in MODIS). Imaging sensors placed on satellites use multiple spectral bands (*e.g.* visible, near-infrared, long-wave infrared, thermal infrared) that allow detection of a wide variety of features, from aerosols and smoke in the atmosphere to chlorophyll in the ocean.

Table B.1. **Components of a space-based remote sensing system and different options**

Component	Issues to be considered	Options
Platform	Type of platform, power and payload	Multiple sensor satellite bus or single sensor satellite? Alternatives to satellite platforms?
	Launch vehicle	Dedicated launch or opportunity mission?
	Orbit type	Near-polar, or geostationary? Sun-synchronous or not?
Sensor	Type of sensor	Active or passive?
	Spectral window	Visible, Infrared (IR), or microwave?
	Type of spatial sampling	Point or imaging?
Data transmission	Type of data coding	Digital or analogue?
	Routing of data	Direct transmission, store and dump, or data relay satellite?
	Access to data	Direct open broadcast or encrypted?
Ground segment	Data receiving stations	–
	Preliminary data processing	Quality control calibration
	Further data processing	Geophysical products, composite products
	Product validation	–
	Data archiving	–
	Data distribution	Data access policy
Data utilisation	Interpretation and analysis	Blend with *in situ* data, synergy with other data, integration into models?
	Scientific exploitation	Direct involvement of users?
	Value added products	For sale or open distribution?
	Operational exploitation	*E.g.* ocean monitoring and forecasting agencies, commercial users?

Source: Adapted from Robinson, 2003.

OECD PUBLICATIONS, 2, rue André-Pascal, 75775 PARIS CEDEX 16
PRINTED IN FRANCE
(03 2008 03 1 P) ISBN 978-92-64-05413-4 – No. 56437 2008